Level A

Investigating
Patterns of Change

Middle School Science & Technology

BSCS *Innovative Science Education*
founded 1958

KENDALL/HUNT PUBLISHING COMPANY
Dubuque, Iowa

This material is based on work supported by the National Science Foundation under Grant No. MDR 8855657. Any opinions, findings, conclusions, or recommendations expressed in this publication are those of the authors and do not necessarily reflect the views of the granting agency.

BSCS Development Team

Rodger W. Bybee, *Principal Investigator* (1988–92)

Janet Carlson Powell, *Project Director* (1988–92)

Kathrine A. Backe, *Staff Associate, Implementation* (1991–92)

Wilbur C. Bergquist, *Staff Associate, Evaluation* (1991–92)

Deirdre Binkley-Jones, *Project Secretary* (1992)

Jan Chatlain Girard, *Art Coordinator* (1989–92)

Sariya Jarasviroj, *Production Assistant* (1992)

Terri Johnston, *Project Secretary* (1991–92)

Donald E. Maxwell, *Staff Associate, Staff Development* (1990–92)

Mary E. McMillan, *Staff Associate, Curriculum Development* (1991–92)

Josina Romero-O'Connell, *Staff Associate, Curriculum Development* (1991–92)

Teresa Powell, *Project Secretary* (1989–92)

Judith Martin Rhode, *Research Assistant*, (1989–91), *Staff Associate, Curriculum Development* (1992)

Joe Ramsey, *Production Assistant* (1992)

William C. Robertson, *Staff Associate, Curriculum Development* (1989–91)

Nancy Smalls, *Project Secretary, Graphics* (1990–92)

Jenny Stricker, *Staff Associate, Curriculum Development* (1992)

Pamela Van Scotter, *Staff Associate, Editing* (1990–92)

Lee B. Welsh, *Production Coordinator* (1989–92)

Yvonne Wise, *Project Secretary* (1989–92)

These titles and dates indicate the primary area of responsibility for each person and the years they worked on the project. Everyone on the project team contributed in numerous ways to create this curriculum.

Photography

Carlye Calvin

NASA

Unicorn Stock Photos—Robert W. Ginn

Biosphere II—Scott McMullen

Richard B. Levine

NAMES Project—Marcel Mirand

Shawn Sigstedt

Unicorn Photos—Joseph L. Fontenct

Dr. Martin Lockley

Charles W. Melton

R. E. Barber

Colorado Springs Utilities

G. M. Hughs Electric

Vanuga Photography

Frances M. Roberts

Mark Schug

Les Van

Student photo on cover—John W. Clark Photography

Artists for First Commercial Edition

Jan Chatlain Girard

Linn and Bob Trochim—Animart

Susan Bartel

Nancy C. Smalls

Janet Huntington-Hammond

Carmen Franco-Stephenson

Bill Ogden—Animation Renderings

PC&F, Inc.—Technical Illustrations

BSCS Administrative Staff

Roger G. Olstad, *Chair, Board of Directors*

Joseph D. McInerney, *Director*

Rodger W. Bybee, *Associate Director*

Larry Satkowiak, *Chief Business Officer*

Board Members

(Continued on p. 357).

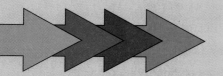

Table of Contents

Preface   ix
Program Overview   xi

UNIT 1

Patterns of Change   1
COOPERATIVE LEARNING OVERVIEW   2

CHAPTER 1

Finding Patterns: Puzzles, Feet, and Music   5

Engage/Explore   **INVESTIGATION:** Puzzlers   6
Explore   **INVESTIGATION:** Driving Day   8
Explore   **INVESTIGATION:** Getting Off on the Right Foot   11
Explain   **READING:** Patterns and Predictions   13
Elaborate   **INVESTIGATION:** Sizes and Swings   16
Elaborate   **READING:** Types of Patterns   21
Elaborate   **CONNECTIONS:** Patterns in Nature   23
Elaborate   **INVESTIGATION:** Moon Watch   24
Elaborate   **INVESTIGATION:** Making Music   25
Evaluate   **CONNECTIONS:** What Do You Know about Patterns?   27

CHAPTER 2

Leafing Out: Patterns That Grow   31

Engage   **CONNECTIONS:** If I Were a Magnolia Tree   32
Explore   **INVESTIGATION:** Beanstalk I   32
Explain   **READING:** What Affects Patterns?   35

Explain	**READING:** Controlled Experiments	37
Elaborate	**INVESTIGATION:** Beanstalk II	39
Elaborate	**READING:** Cause and Effect in My Life	41
Elaborate	**CONNECTIONS:** Captain Sneezy's Cold Cure	43
Elaborate/Evaluate	**INVESTIGATION:** Beanstalk III or Radishes on the Rise	46
Evaluate	**CONNECTIONS:** Plants All Over	47

CHAPTER **3**

Predictions: More Than Just a Guess 49

Engage/Explore	**INVESTIGATION:** Finney's Funny Food	50
Explore	**INVESTIGATION:** The Power of Attraction	52
Explain	**READING:** Making Accurate Predictions	55
Elaborate	**INVESTIGATION:** Will It Sink or Float?	61
Evaluate	**CONNECTIONS:** Minding Your Ps (Predictions) and Qs (Quality and Quantity)	63

CHAPTER **4**

The Moon and Scientific Explanations 65

Engage/Explore	**READING:** Moon Legends—Another Way of Explaining Patterns	66
Explore	**CONNECTIONS:** Tell It Your Way	68
Explore	**INVESTIGATION:** Moon Movies	68
Explore	**CONNECTIONS:** Moon Movies and Predictions	71
Explain	**READING:** What Makes an Explanation Scientific?	72
Explain/Elaborate	**INVESTIGATION:** Explaining Phases	72
Evaluate	**CONNECTIONS:** The View from Earth	75

CHAPTER **5**

Recognizing Patterns of Change 79

Evaluate	**READING:** Can You Recognize Patterns of Change?	80
Evaluate	**CONNECTIONS:** Putting the Pieces Together	81
Evaluate	**INVESTIGATION:** What's Going On Here?	82
Evaluate	**READING:** The Value of Patterns and Scientific Explanations	84
Evaluate	**CONNECTIONS:** Sorting Out the Patterns	86

UNIT 2

Explanations For the Patterns of Change on the Earth 91
COOPERATIVE LEARNING OVERVIEW 92

CHAPTER 6

Scientific Explanations Begin with a Question 95

Engage/Explore **INVESTIGATION:** How Do You Know? 96
Explore **INVESTIGATION:** Think Like a Cube! 98
Explain **READING:** Science Is a Way of Explaining 101
Elaborate **INVESTIGATION:** The If–Then Box 104
Evaluate **CONNECTIONS:** What Did We Do? 107

CHAPTER 7

Volcanoes, Earthquakes, and Explanations 109

Engage **CONNECTIONS:** Can You Imagine? 110
Explore **INVESTIGATION:** And Along the Way They Met . . . 110
Explain **READING:** Can You Believe It? 131
Elaborate **INVESTIGATION:** Patterns on the Earth 133
Evaluate **CONNECTIONS:** Can You Explain the Observations? 138

CHAPTER 8

Connecting the Evidence 141

Engage **INVESTIGATION:** Desks on the Move 142
Explore **INVESTIGATION:** Where Have the Continents Been, and Where Are They Going? 143
Explain **READING:** Combining Ideas 144
Explain/Elaborate **CONNECTIONS:** Putting It Together 148
Explain/Elaborate **INVESTIGATION:** But Why Should They Move? 148

Elaborate	**INVESTIGATION:** Near the Edges	152
Elaborate/Evaluate	**INVESTIGATION:** Plate Tectonics Research	156
Evaluate	**CONNECTIONS:** Looking Back	161
Evaluate	**CONNECTIONS:** Listen to How They Say It	161

CHAPTER 9

Using Scientific Explanations   163

Explore/Explain	**READING:** Scientific Explanations for Experts	164
Elaborate	**INVESTIGATION:** Plate Tectonics in Outer Space	165
Evaluate	**INVESTIGATION:** The Mystery Planet	170

UNIT 3

Responding to Patterns of Change   173

COOPERATIVE LEARNING OVERVIEW   174

CHAPTER 10

What Causes Weather Patterns?   177

Engage/Explore	**INVESTIGATION:** Water on the Move	178
Explore	**INVESTIGATION:** Wind in a Box	179
Explain	**READING:** Changing Weather Patterns	182
Explain/Elaborate	**INVESTIGATION:** Winds above a Rotating Earth	188
Elaborate	**CONNECTIONS:** Picture This	192
Evaluate	**CONNECTIONS:** Science in Your Bathroom	193

CHAPTER 11

Natural Events and Natural Disasters   195

Engage	**CONNECTIONS:** When You See This, What Do You Think?	196
Explore	**INVESTIGATION:** Miniature Events	200
Explain	**INVESTIGATION:** Presenting Events	201

Explain/Elaborate	**CONNECTIONS:** You Decide	215
Elaborate	**INVESTIGATION:** Twisters and Numbers	217
Elaborate/Evaluate	**INVESTIGATION:** Really Stormy Weather	219
Evaluate	**CONNECTIONS:** People and Natural Events	225

CHAPTER 12

Making Decisions and Solving Problems 227

Explore	**INVESTIGATION:** Standing Against the Wind	228
Explore	**INVESTIGATION:** The House on the Windy Plain	230
Explain	**READING:** A Process for Solving Problems	233
Explain	**CONNECTIONS:** The Popcorn Cube	234
Explain	**READING:** Decisions Are Part of the Process	234
Elaborate/Evaluate	**INVESTIGATION:** How Did This Thing Get Here?	236
Elaborate/Evaluate	**CONNECTIONS:** Balanced Decisions	238

CHAPTER 13

You're Probably Right 241

Engage	**CONNECTIONS:** Those Difficult Decisions	242
Explore	**INVESTIGATION:** It's in the Bag	242
Explain	**READING:** Definitely, Probably, Maybe	247
Elaborate	**INVESTIGATION:** What Are the Chances?	249
Elaborate	**INVESTIGATION:** What Will Happen Here?	251
Evaluate	**CONNECTIONS:** Do You Agree?	253
Evaluate	**CONNECTIONS:** You Be the Judge	254

UNIT 4

Patterns and People 257
COOPERATIVE LEARNING OVERVIEW 258

CHAPTER 14

It's Everywhere 261

Engage/Explore	**CONNECTIONS:** Look at the Evidence	262
Explore	**INVESTIGATION:** All in a Day's Garbage	264

Explain	**READING:** The Garbage Crisis	270
Explain	**CONNECTIONS:** Through the Years	277
Explain/Elaborate	**INVESTIGATION:** Pitch and Take	278
Elaborate	**INVESTIGATION:** Everything in Its Place	280
Evaluate	**CONNECTION:** Down by the River and Out in the Desert	282

CHAPTER 15

Solving Problems   287

Explore	**CONNECTIONS:** On Almost Every Corner	288
Explore/Explain	**INVESTIGATION:** The Choice Is Yours—Projects, Part I	289
Elaborate/Evaluate	**INVESTIGATION:** Presentations—Projects, Part II	294
Evaluate	**CONNECTIONS:** St. Louis and San Diego: Opposite Decisions?	296

How To #1	301
How To #2	305
How To #3	309
How To #4	317
How To #5	323
How To #6	325
How To #7	331
How To #8	333
How To #9	341
How To #10	345

Glossary	351
Acknowledgments	367
Index	369

Preface

Welcome to *Middle School Science & Technology!* We developed this science program specifically for middle school students, such as yourself. In our process of development, we considered middle school students first, then middle school teachers, and then other aspects of middle schools. To develop a program that reflects what middle school students want and need, we had to communicate with many middle school students. We ended up talking with or writing to more than 20,000 students across the United States and in South America. If you look at the credits on p. 367 in the back of the book, you can see where all those students went to school.

We took the following steps to find out what middle school students liked or did not like about science and technology and what we therefore should include in this program:

- We interviewed students about their preferences for science topics. We wanted to know which topics they liked and which they did not. We found out that many topics appealed to middle school students, so we included a great variety of topics in this curriculum.

- We surveyed students to find out which teaching methods they liked and which they did not, because we found out that the preference students had for particular topics depended on how their teacher presented the topic. We found out that very few students liked lectures but that everyone liked *doing* activities. As a result this program depends on your doing and thinking about many activities, and your teacher will be lecturing only a very few times.

- Then we wrote the first draft of the book and let teachers and students try it out for a school year. This trial period is called a field test. As they tried the program, we visited their schools to find out which activities and ideas were working and which were not. Also the field-test teachers and students wrote to us to keep us up-to-date on their progress.

- Using that feedback from middle school students and teachers, we wrote our second draft of the book. More students and teachers then tried it out for another year. This was the second field test. Once again the field-test participants let us know which activities and ideas were going well and which were not.

- Finally, we compiled all of these comments and reactions to create the book you are about to begin using. This book has combined the input of thousands of students and teachers who took the time to tell us how to improve this book so that you could use and enjoy it.

We hope *Investigating Patterns of Change* is a program that makes science and technology enjoyable for you to learn. We also hope the activities and ideas in this program make you want to learn more about science and technology. Before long we will be working on the second edition of this program. The second edition will be a revised version of this book. The revisions we make will be based on the feedback of the students and teachers who use this book. If you have comments that you think will improve the second edition for other middle school students, please send your comments to the following address:

BSCS
Attention: MSP
830 N. Tejon Street, Suite 405
Colorado Springs, CO 80903

Sincerely,

Rodger W. Bybee
Principal Investigator

Janet Carlson Powell
Project Director

Program Overview

Middle School Science & Technology is probably unlike other science programs you have used in school. This overview briefly describes some of the key features of the program. You will learn more about these special features of the program when you complete the introductory unit.

Themes

The organization of information in this program might be different from other science programs that you have seen or used. This is because we have organized each level around a unifying theme. One way of describing a unifying theme is as a common idea that keeps coming up to tie other ideas together. Writers use a unifying theme, called a story line, when they write novels. Often music will have a unifying theme that you hear over and over in the piece of music; Beethoven's Fifth Symphony is a famous example of this.

We used a different unifying theme at each level, but each theme has the same purpose: to provide a thread that ties together many scientific and technological ideas. To help keep the unifying theme from becoming repetitive or boring, we use a different curriculum emphasis and focus question for each unit. The curriculum emphasis helped us decide what part of science to emphasize in each unit. The focus question provided a way to connect the unifying theme and the curriculum emphasis to the unit topics in a meaningful way for middle school students. The scope and sequence chart on the next page shows how we organized all of these parts to form an entire program that extends over 3 years. You can use this chart to see how what you are learning this year relates to what is coming in Level B and Level C. Look closely at the focus questions. These questions should give you a sense of the direction your work in each unit will take.

The Five "Es"

As you begin using your textbook, you will notice that each page has one or more of the following words at the bottom: *Engage, Explore, Explain, Elaborate, Evaluate.* These words make up five phases that you will use as you learn an idea. These phases describe what you are doing as a learner and what your teacher is doing as a teacher. You will go through each phase approximately once each chapter. For example, when you are doing an **engage**

Scope and Sequence

Level A: Patterns of Change

Unit	1	2	3	4
Curriculum Emphasis	Personal dimensions of science and technology	The nature of scientific explanations	Technological problem solving	Science and technology in society
Focus Question	How does my world change?	How do we explain patterns of change on the earth?	How do we adjust to patterns of change?	How can we change patterns?

Level B: Diversity and Limits

Unit	1	2	3	4
Curriculum Emphasis	Personal dimensions of science and technology	Technological problem solving	The nature of scientific explanations	Science and technology in society
Focus Question	What is normal?	How does technology account for my limits?	Why are things different?	Why are we different?

Level C: Systems and Change

Unit	1	2	3	4
Sub-theme	Systems in Balance	Change through time	Energy in Systems	Populations
Curriculum Emphasis	Personal dimensions of science and technology	The nature of scientific explanations	Technological problem solving	Science and technology in society
Focus Question	How much can things change and still stay the same?	How do things change through time?	How can we improve our use of energy?	What are the limits to growth?

activity, you will be thinking about a new idea. Then you will **explore** that idea in one or several activities. Next you will either **explain** your understanding of the idea, listen to the teacher **explain** more about an idea, or a combination of the two. Then you will **elaborate** your understanding of the idea, usually by doing another activity. Finally you and your teacher will **evaluate** your understanding of the idea. If your evaluation shows that you are successful in understanding the idea, it is time to be engaged in a new idea and to go through these phases for that idea.

Cooperative Learning

We have incorporated cooperative learning strategies into about two-thirds of the program for a variety of reasons. One reason is that cooperative learning gives you a chance to learn and practice how to work successfully with others. The skills that you will gain will become more important to you as you begin to work in an employment setting. Also researchers have shown that cooperative learning can increase the success of students in science class. Cooperative learning also cuts down on the amount of materials needed for each investigation. Your school then can afford to buy other materials so that you can do more investigations. In addition including cooperative learning strategies helps you work like professional scientists and engineers, who do most of their work in cooperative settings in laboratories.

The Characters

We use four cartoon characters in this book (see page xiv). Al, Marie, Isaac, and Rosalind are in the book for four main reasons:

1. To provide a concrete method for demonstrating the value of different learning styles,
2. To make the book friendlier,
3. To teach some of the history of science, and
4. To provide a type of positive role model.

The characters are introduced and described thoroughly in the introductory unit. That information identifies the strengths of each character's learning style, the historical reference for the character's name, and a bit about the character's personality. In the text the characters provide examples of why science is something everyone can do, because they are a diverse group of learners and each contributes something positive to the group.

Questions

There are two primary places you will find questions in this book: in readings as stop and discuss points, and at the end of investigations as Wrap Ups. Some of the students who field tested the curriculum commented that the questions were hard to answer because they had to think about their answers instead of copying

Al

Isaac

Marie

Rosalind

them out of the book. This reaction satisfied one of our goals for this program: to increase your ability to think critically. You will notice that many of the questions in the book have more than one answer that can be considered to be correct. This is because we tried to write questions that you can answer in a variety of ways as long as you provide support for your answer. If this is a new way of answering questions for you, you might feel frustrated at first, but eventually you might find that you enjoy learning this way. Using this questioning method places you more in charge of your own education.

Assessment

Because the topics, themes, and questions in this book are different from most other science programs, it only makes sense that we would include different assessment methods for measuring your success and progress. In many programs you are assessed only by your performance on quizzes and tests. In this program we have recommended that teachers use a variety of assessment strategies that include daily notebooks, checklists, performance tests, short-answer tests, and portfolios to measure how much you have learned and improved and to identify areas that you might want to focus on for future improvement. So don't be surprised if you have "tests" that don't remind you of the tests you are used to taking.

Safety

As in any science program, safety is a concern for everyone who uses *Middle School Science & Technology*. We have made every effort to alert you, the learner, to potentially dangerous situations or materials. We have marked these places with the following symbol:

▲ CAUTION:

In addition your teacher should tell you about the safe behaviors that you should use in a science classroom. It is your responsibility to follow all safety warnings, rules, and procedures to avoid possible injury to yourself or others.

Patterns of Change

Have you ever watched a mystery movie? Perhaps you noticed that the detective in the movie recognized small details you missed. And as you watched the movie, the puzzle began to fit together. The detective probably looked for a pattern and then noticed how it changed.

Although you may not know it, life is full of puzzles that you can solve by noticing patterns. Some of the people who specialize in puzzles and patterns are detectives who work in police departments. But another group of detectives exists: scientists. Scientists observe puzzles in nature and propose solutions to those puzzles. Often when scientists find that something puzzles them, it is because a pattern has changed. Noticing *how* a pattern changes often provides a solution to a puzzle.

You might not think that you know much about patterns, but you probably are more of a scientist than you realize. Science begins with keen observations that can help unravel mysteries. In this unit you will have opportunities to do just that—to look at things carefully and to unravel mysteries.

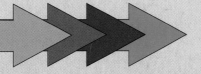

COOPERATIVE LEARNING OVERVIEW

Before you begin working on Chapter 1, take time to familiarize yourself with the social skill for Unit 1. Throughout this entire unit, you will be practicing the social skill, express your thoughts and ideas aloud. When you practice this social skill, use this social skill rule: listen politely to others. Try to think of using them together in a combined social skill: express your thoughts and ideas aloud and listen politely to others.

Study the character scene on the facing page. Some of you probably have had different experiences with cooperative learning. You might have had a lot of practice in previous grades. Or this might be your first experience with cooperative learning. Either way, you will have the opportunity to work in cooperative groups frequently throughout this book. Remember to refer to the cooperative job descriptions as often as necessary until you are sure about the duties of each job. Always take time to think about a new skill when we first introduce it.

Right now, as Al said, you need to create a T-chart for this social skill. Do so according to your teacher's directions. Keep this T-chart handy. Even though activities will mention other social skills, you need to keep practicing your unit skill through the entire unit.

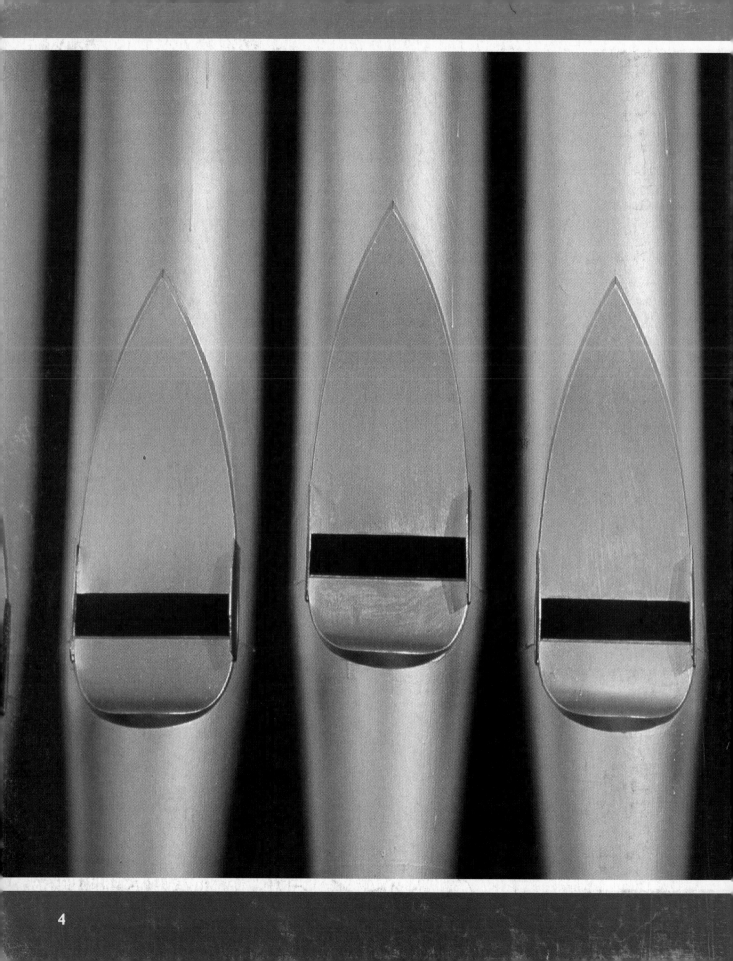

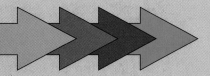

Finding Patterns: Puzzles, Feet, and Music

Do you know what the weather will be like tomorrow? Do you know what subjects you will study in school next Tuesday?

Sometimes you have a good idea about what is going to happen, and sometimes you only can guess. You can be fairly sure about what subjects you will study in school next Tuesday because you are familiar with your school schedule. When you are familiar with something, you don't have to guess about what will happen next.

INVESTIGATION:
Puzzlers

Many times scientists investigate nature the same way we solve puzzles. They have to look for clues to help them figure out what is happening. How good are you at solving puzzles? Do you ever use a code to communicate? In this activity you will solve a puzzle and figure out a code.

Materials

For each team of two students:
- 2 copies of the Decoding Chart
- glue, stapler, or tape

Procedure: Part A—The Social Skill

1. Get into your teams.
2. Make your notebook entry.
3. In your notebook write the first name and last name of your teammate.
4. Individually make a list of two reasons why using each other's names is important in teamwork.
5. Share and discuss your lists.

Procedure: Part B—Mystery Language Puzzle

1. Pick up two copies of the Decoding Chart.
   This is the Manager's job.
2. Look at the Mystery Language Puzzle with your partner.

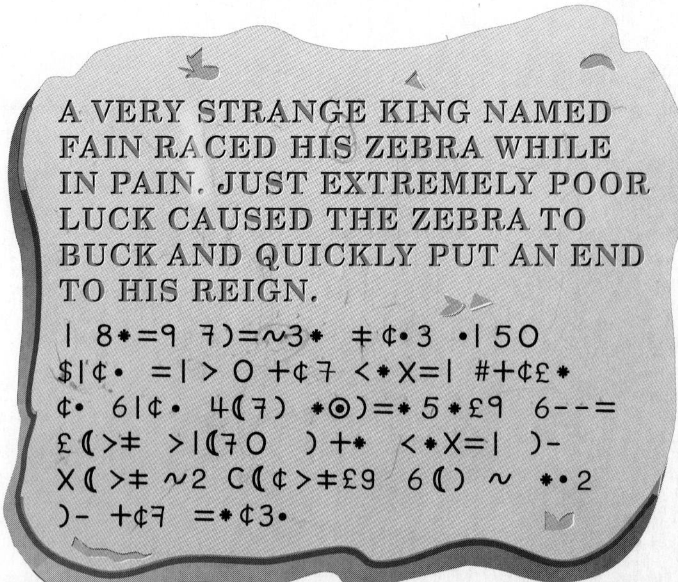

Working Environment

You will work in cooperative teams of two. All people in the group should follow the Team Member role. In addition to that role, one of you will be the Manager and the other will be the Communicator, according to your teacher's directions. Check the job descriptions of each person. Move your desks side by side or sit together at a table. Work on the social skill Use your teammate's name.

Figure 1.1

MYSTERY LANGUAGE PUZZLE—This diagram shows the same words written in English and in the mystery language.

A VERY STRANGE KING NAMED FAIN RACED HIS ZEBRA WHILE IN PAIN. JUST EXTREMELY POOR LUCK CAUSED THE ZEBRA TO BUCK AND QUICKLY PUT AN END TO HIS REIGN.

3. Read the decoded letters aloud from the Mystery Language Puzzle while your partner records them on the blank Decoding Chart.

   For example, the first symbol is the number 1, which stands for the letter A. Try having the Communicator read letters while the Manager records them.

4. Check your team's Decoding Chart.

   It should now show what the mystery language alphabet is.

5. Copy your partner's chart if you do not have one.

   Both partners should have a copy of a completed Decoding Chart. Check your chart to be sure you can find the symbol for each letter.

6. Put your chart into your notebook.

   Use tape, staples, or glue to attach the chart to a page in your notebook.

7. Ask the teacher for a new message to decode.

   This is the Communicator's job.

8. Enter the answers to your messages into your notebook.

Procedure: Part C—Short Puzzles

1. In your notebook, copy Puzzles A through D, each on a separate line.

   See Figure 1.2.

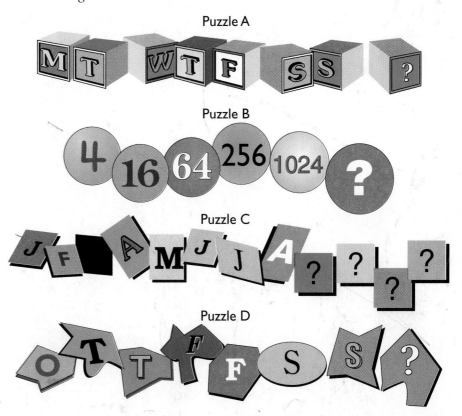

Figure 1.2

Short puzzles A, B, C, D. See if you can solve each one.

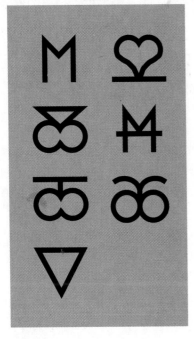

Puzzle E

Figure 1.3

Here is a puzzle that might be easier to solve if you could first cut the symbols in half.

2. Look at the puzzles. As a team decide what goes in each blank.

3. After you and your teammate have agreed on answers for Puzzles A through D, record the answers individually in your notebooks.

   STOP: Are you using each other's names? Are you listening politely to each other?

4. Look at Puzzle E.

   See Figure 1.3.

5. Decide what the last symbol should be and draw it in your notebook.

Wrap Up

Write the words "Wrap Up" in your notebook. Discuss the following questions with your partner and write your answer for each question in your notebook. Be sure each of you can explain your answers during a class discussion.

1. In the Mystery Language Puzzle from Part B, what was the first symbol you were able to decode?

2. When your team decoded the first set of symbols, how did you determine the answers?

3. Imagine that you meet a friend in the hall and that he or she wants to know what you just did in science class. What would you tell your friend?

4. What is your teammate's name?

5. How well did you use your teammate's name?

6. How did using names help you complete this investigation?

INVESTIGATION:
Driving Day

Every birthday is a special event, but one particular birthday is extra special. That's the day you become eligible for a driver's license. In this investigation you will have a new puzzle to solve—figuring out on which day of the week you will turn 16 years old.

Figure 1.4

This is an example of what
your notebook entry might
look like.

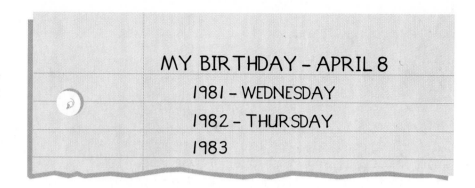

MY BIRTHDAY – APRIL 8

1981 – WEDNESDAY

1982 – THURSDAY

1983

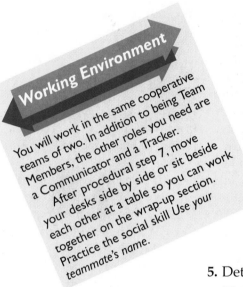

Working Environment

You will work in the same cooperative teams of two. In addition to being Team Members, the other roles you need are a Communicator and a Tracker. After procedural step 7, move your desks side by side or sit beside each other at a table so you can work together on the wrap-up section. Practice the social skill Use your teammate's name.

Materials

For each student:
- monthly calendars

Procedure

1. Make your notebook entry.

2. Find the calendar page that shows the year, month, and day of your birth.

3. Determine on which day of the week you were born.

4. Record in your notebook the day you were born. *See Figure 1.4.*

5. Determine the day of the week you turned 1 year old.

    Notebook entry: Record this day in your notebook.

6. Do the same for ages 2 and 3.

    You also should have recorded in your notebook the day of the week you turned 2 and the day of the week you turned 3.

7. Figure out on which day of the week you will be eligible to drive.

    This will not be as easy as the previous step because you do not have a calendar for that year. Figure out a way to tell without looking up the day.

Wrap Up

After moving your desk next to your partner's, write the words "Wrap Up" in your notebook. Share your thoughts and ideas with your partner. In your notebook write your answer for each question. Be sure each of you can explain your answers during a class discussion.

1. On which day of the week will you be eligible to drive?

2. On which day of the week will your partner be eligible to drive?

Sure hope your driving birthday doesn't fall on Friday, the thirteenth!

Measuring Time

Imagine that you're trying to make a date with a friend to go see a movie. How do you describe the time when you should meet? What if you could not mention the day or the time because the words to describe those things were not part of your language? Perhaps you could tell your friend to count the number of sunrises and, on the day of the fourth sunrise, wait until the bats fly from their cave in the evening and meet you then. Sure sounds complicated, doesn't it?

Before calendars, people measured time by watching motion—the motion of the earth, moon, sun, and stars. Ancient people based their system of time keeping on some regularly occurring event, such as the time from one full moon to the next or the time from one sunrise to the next.

You probably take the yearly calendars you just used in Driving Day for granted. It might surprise you to know that the calendar has gone through many changes and improvements since someone first got the idea of organizing time. People based the first calendars on the pattern of change in the shape of the moon and divided the year into 12 months of 30 days each, or 360 days. Because it actually takes 365 and $\frac{1}{4}$ days to get from one spring to the next, these early calendars became less accurate each year. To keep things from getting totally out of step, people had to keep adding extra days. The process of tracking time became very complicated.

Eventually the Egyptians stopped looking to the moon and chose Sirius, the Dog Star, to mark their year. Once a year Sirius rose in the morning in direct line with the rising sun. The Egyptians also found that 12 months of 30 days each could provide a useful calendar of the seasons if they added another 5 days at the end, to make a year of 365 days. This time, the difference between the calendar year and the "solar" year was so small that it took many years, far longer than any one person's lifetime, for the error to disturb daily life. So this was

What did they do about people's birthdays during those lost 10 days? Did they just skip 'em?

the calendar that Julius Caesar adopted—it was called the Julian calendar.

We would probably be using the Julian calendar today if it were not for the actions of Pope Gregory XIII in 1582. Even though the Julian calendar added an extra day to every fourth year, the calendar was still falling behind. The difference between the Julian calendar and an actual solar year is only 11 minutes and 14 seconds per year. That adds up to 8 days every 1,000 years. By 1582 the calendar did not match with the seasons and seasonal celebrations, such as Easter. Pope Gregory XIII decided to change the calendar so that it reflected true solar years. The Pope based the changes on the ideas of several astronomers.

First the Pope decreed that the day following October 4, 1582, would be October 15. He also declared that people would skip leap year (every fourth year to which 1 day was added) once every 128 years. Further, for the years that end a century, only those that are divisible by 400 should be leap years. So the years 1800, 1900, and 2100 have no extra day, but the year 2000 does have the extra day. This shortened the calendar year and made it almost exactly the same length as the solar year. In fact the difference is less than $\frac{1}{2}$ a minute per year, and the calendar will get out of step only 1 day every 3,000 years.

3. Could you determine this day accurately without looking at any calendars? Explain your answer.

4. Describe how you were able to determine on which day of the week you will be eligible to drive.

5. As you discussed these questions, did you remember to use each other's names?

 INVESTIGATION:
Getting Off on the Right Foot

Perhaps you know your shoe size, but how *long* do you think your foot is? Do you suppose that everyone in your class has the same foot length? In the previous two investigations, you predicted things based on puzzles and calendars. You will end this investigation by predicting foot sizes based on measurements you will make.

Materials

For each student:
   ◼ 1 piece of light-colored construction paper
For each team of two students:
   ◼ 1 metric ruler or tape measure
   ◼ masking tape, one 10-cm strip
   ◼ 1 wide, felt-tipped, nonpermanent marker

Procedure: Part A—The Social Skill

1. Create a T-chart in your notebook.

   Title it, "Move into your groups quickly and quietly." Label the left column "Sounds Like" and the right column "Looks Like."

2. Fill in the columns according to the social skill.

   As you discuss with your teammate and fill in the columns, consider the following:

   ◼ *What kind of noise is appropriate as you move into groups?*

   ◼ *What kind of noise is inappropriate as you move into groups?*

   ◼ *How quickly is "quick"?*

   ◼ *What kind of behavior is appropriate and inappropriate as you move into your groups?*

3. When your teacher says, "Go," see which team can move into its group the most quickly and quietly.

Working Environment

You will work in cooperative teams of two. Each of you will be a Team Member. In addition, one of you will be the Communicator, and the other will be the Manager and Tracker. Check the job descriptions to make sure you understand your roles. Work on the social skill Move into your groups quickly and quietly.

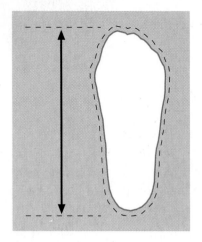

Figure 1.5

Measure your foot tracing from your big toe to the end of your heel as shown.

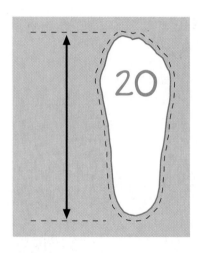

Figure 1.6

Your finished tracing might look something like this.

Procedure: Part B—The Activity

1. Collect the necessary materials.

   Remember, it is the Manager's job to collect the materials.

2. Take off your right shoe.

   Keep your sock on.

3. Have your partner trace your foot.

   Have one person stand with his or her right foot on one of the pieces of construction paper while the other person traces the outline of the foot. While your partner is tracing your foot, your entire foot should be on the construction paper.

4. Repeat step 3 for your partner.

   Put your shoe back on.

5. Use a metric tape measure or ruler and measure the length of your foot tracing in centimeters.

   See Figure 1.5.

6. Round off your measurement if it is not a whole number.

   If you do not know how to round off, go to How to Round Off Numbers (How To #1).

   STOP: Now is a good time to help each other round off numbers by expressing your thoughts and ideas aloud to your partner.

7. Write this number in the center of your tracing.

   See Figure 1.6.

8. Take your foot tracing to the graph that your teacher has on the wall or chalkboard and tape it above the number that corresponds to your foot length.

9. Draw a picture in your notebook, including the shape of your class graph and its two labeled axes.

   If you want to know more about graphs, read How to Identify the Parts of a Graph (How To #2).

Wrap Up

Write the words "Wrap Up" in your notebook. Discuss the following questions with your partner and write your answer to each question in your notebook. Be sure you could explain your answers if your teacher called on you.

1. Look at the graph and describe where your tracing is, compared with where your classmates' tracings are.

2. How many students have foot sizes of 20 cm?

3. How many students have foot sizes of 25 cm?

4. Write a summary of this graph. Be sure your summary includes:
   a. how many people were measured,
   b. the smallest and largest foot size,
   c. the most common foot size, and
   d. a description of the shape of the graph.

5. If another class of middle school students constructed a graph of their foot sizes, would the graph have the same shape?

6. Imagine that you invited a professional football team into your classroom. Suppose the football players took off their shoes, outlined their feet, and added their information to your class graph. Would the shape of the graph change, and if so, how?

7. How might the shape of your original class graph change if you measured a group of preschool children's feet and added their foot measurements to the graph instead of the football team's measurements?

8. How could you and your partner improve how quickly and quietly you move into your group?

9. When were you the best at moving quickly and quietly—at the beginning of the investigation or at the end?

READING:
Patterns and Predictions

In the three previous investigations, you solved puzzles, discovered on which day of the week you would be eligible to drive, and discussed the shape of a graph of foot lengths. You looked at a variety of things. You might have wondered what these activities had in common with each other. To solve the questions in each investigation, you had to determine a **pattern.** A pattern is a collection of things or events that repeat themselves. Because we know what to expect next from a certain pattern, we can make predictions.

There are many kinds of patterns in the world. For example, most of us know that to cross a street safely, we must watch for cars. Traffic lights may cause the pattern to change. You might notice a rush of cars, then no cars. If you want to cross safely, you wait until the right moment in the pattern.

Recall the sequence of letters O T T F F S S ? . The sequence doesn't seem to make much sense until you suddenly see the pattern that the letters represent. They are the first letters of words for the numerals one through seven. Using this pattern, it's easy to predict what comes next.

Figure 1.7

Your class graph probably looked something like this.

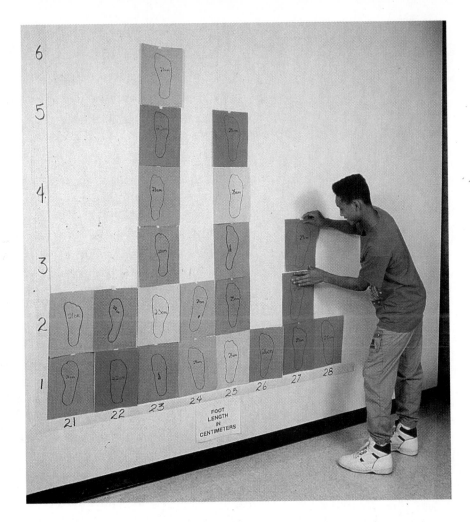

Think back to another example. After you and your classmates taped your foot tracings on the board, your completed graph had a particular pattern: lots of foot tracings near the middle and fewer near the ends. Its general pattern was similar to the shape of the graph in Figure 1.7.

Stop and Discuss

1. Now that you have seen several examples, describe how we can use patterns.

2. Try making this prediction: What will most of your classmates be doing tomorrow at 4:00 A.M.?

3. What pattern did you use to make your prediction?

If you know a pattern, you can use it to make **predictions**. A prediction is a statement about the future, based on information. Patterns provide one type of information. Throughout this year you will study patterns of change and use them to make predictions.

You have seen several examples of puzzles with patterns that we made up. There also are puzzles with patterns in the world around you. Some patterns are unusual for us, and we might not recognize them immediately. In 1799, for example, a construction worker in the French army found a most unusual piece of stone near Rosetta, Egypt (see Figure 1.8). It had writing carved into it, so he knew it had a pattern. But he could not read the writing. He thought it was interesting, however, because three different patterns of writing were on the stone. Scholars recognized that one of the patterns was ancient Greek and the other two were ancient Egyptian forms of writing. Because these scholars could already read the Greek, they were able to decode the Egyptian writings, which they never before had been able to do. The scholars realized that the two Egyptian forms were closely related. One form, called hieroglyphics, was the sacred form, and only religious officials

Figure 1.8

The Rosetta stone is inscribed with three different languages. The top inscription is in hieroglyphics. The middle inscription is in demotic. Both of these are Egyptian languages. The bottom inscription is in Greek. Each inscription repeats the same phrases honoring a Greek king of Egypt called Ptolemy V. The Rosetta stone is kept at the British Museum.

The Bettmann Archive

used it. The word hieroglyphics comes from the Greek words meaning "sacred carvings." The other form, called demotic, was the common, everyday form. The scholars named this stone tablet the Rosetta stone. Because the stone provided a relationship between the three languages, scholars could decode other Egyptian carvings on other stones, as well as this stone. When scholars decoded the stones, they unlocked years of history. None of this information would have been known unless someone had noticed the pattern. The decoding of that stone provided information about everything from the crowning of kings to medical history.

We can use patterns to make predictions rather than guesses. The ability to use patterns to make predictions is an important skill. We expect doctors to recognize patterns of illness so that they can predict what actions will cure the illness. Auto mechanics can make predictions about what is necessary to repair a car by recognizing patterns in the noises an engine makes. You can predict how your parents will react to a particular piece of news because you know the pattern of how they have reacted in the past.

Of course, recognizing patterns isn't always easy, and the predictions you make from patterns aren't always correct. The more you look *at* and look *for* patterns, the better you'll be at recognizing them and putting them to use. The following activities will help you do just that.

INVESTIGATION:
Sizes and Swings

In the previous reading you read about some everyday patterns such as traffic and some unusual patterns such as the Rosetta stone from Egypt. Sometimes patterns are familiar, and sometimes they are new to us. In the investigation that follows, your challenge will be to identify the patterns.

Working Environment

You will work in cooperative teams of two at three different stations. When you arrive at a station, make a notebook entry that includes the number of the station. You both will be Team Members. In addition, one of you will be the Communicator, and the other will be the Tracker. Continue to practice the skill Move into your groups quickly and quietly.

Materials

For each team of two students:
For Station 1:
- 2 sheets of Graph Paper with Axes Set Up
- Foot Length Data Table
- 1 metric ruler

For Station 2:
- 1 metric measuring tape

For Station 3:
- 1 pendulum setup
- 1 watch or clock with a second hand

Explain ■ *Elaborate*

	AGE	FOOT LENGTH (cm)
Child	0–6	
Adolescent	10–14	
Adult	18+	

Figure 1.9

Figure 1.9

This is a sample of the chart you will use to record the foot measurements. Copy this into your notebook.

Procedure for Station 1

1. Copy the chart in Figure 1.9 into your notebook.

   Leave plenty of space (about half a page).

2. Fill in the chart you have made by writing down foot length measurements.

   If someone's foot length is 14 cm and he or she is 3 years old, you would write down the foot length in the first row of the chart in the child category. See the examples in Figure 1.10.

3. Use Xs to graph the data on children's foot measurements.

   Use the top section of the prepared graph paper at this station and refer to How to Plot Data on a Graph (How To #3) if necessary.

4. Use Xs to graph the data on adolescent foot measurements.

   Use the middle section of the graph paper.

5. Use Xs to graph the data on adult foot measurements.

   Use the bottom section of the graph paper.

6. Compare the three graphs you have made.

7. Record in your notebook any pattern you notice about the graphs. Answer the wrap-up questions and then go to another station.

Figure 1.10

As you fill in your chart, it will begin to look like this. The first person on your data table is 1 year old and has a foot length of 10 cm. So you would write 10 in the first row.

	AGE	FOOT LENGTH (cm)
Child	0–6	10, 14, 18
Adolescent	10–14	24, 20, 22
Adult	18+	26, 29, 26

Wrap Up for Station 1

Discuss the wrap-up questions with your partner and record your answers in your notebook before you go on to another station.

1. Describe the difference between the foot measurement of a child and that of an adolescent.

2. Describe the difference between the foot measurement of an adolescent and that of an adult.

3. Compare the three graphs you have made and describe the patterns you see.

4. Based on the numbers you have, predict a common foot measurement for 70-year-old people. Explain your answer.

Figure 1.11

To measure your height, stand with your back against the wall and have your partner place a book on your head. Measure to the bottom edge of the book. Stand up straight!

Elaborate

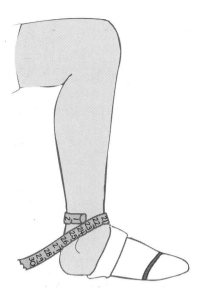

Procedure for Station 2

1. With your partner make and record the measurements listed below.

   If you are not certain how to do some of the measurements, see the accompanying illustrations. Remember to record the number and units for each measurement beside the list. For example, foot length = 24 cm.

   - height
   - ankle circumference
   - thumb circumference
   - length of forearm (elbow to wrist)
   - length of ear
   - foot length (if you don't already have it)
   - length of little finger
   - arm span

Figure 1.12

To measure your ankle circumference, roll your sock down and measure around the bony part—the largest part of your ankle.

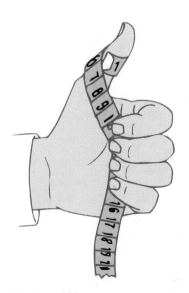

Figure 1.14

Your partner will help you measure your arm span. Stand with your arms straight out at each side at shoulder height. Your partner should measure from the tip of one middle finger across your back to the tip of your other middle finger. You may need to ask another team for help.

Figure 1.13

To measure your thumb circumference, measure around the knuckle of your thumb.

Wrap Up for Station 2

Discuss this question with your partner and record your answer in your notebook before going on to the next station.

1. What patterns did you find in your measurements?

   Here's one hint: The length of your ear and your thumb circumference are about the same.

Procedure for Station 3

1. Make sure the pendulums are still.

   There are two weights hanging on a string. These are called pendulums.

2. Start one pendulum swinging.

   Do not hold the second pendulum.

3. Stop the pendulum after 5 seconds.

   Tracker: Remember to keep track of the time.

4. Record what you observed.

   Notebook entry: Record your observations.

5. Predict what would happen if you let the pendulum swing for 60 seconds.

   Notebook entry: Record your prediction.

6. Test your prediction.

   Notebook entry: Record your observations.

7. Predict what would happen if you let the pendulum swing for 3 minutes.

   Notebook entry: Record your prediction.

8. Test your prediction.

   Notebook entry: Record your observations.

Wrap Up for Station 3

Discuss these questions with your partner and record your answers in your notebook before going on to the next station.

1. Describe the pattern of change you saw in the pendulums.
2. How accurate were your predictions for the patterns of change you observed in the pendulums?
3. Create an explanation for the patterns you saw in the pendulums' motion.

Wrap Up for the Social Skill

With your partner, complete this Wrap Up after you have completed the work at all three stations. Record your answers.

1. On a scale of 1-10, 10 being the best, rate your team for how well you are expressing thoughts and ideas and listening politely to each other.
2. On a scale of 1-10, rate your team for how quickly and quietly you moved among stations.

READING:
Types of Patterns

In the investigation Sizes and Swings you found different kinds of patterns. At Station 1 you identified a pattern in the way people's feet grow. People's feet get longer until their feet stop growing. A **trend** is a type of pattern that goes in a particular direction. For example, if you examine the pattern of human growth over time, you would see these three trends: height increases from birth until about age 20, height stays the same from age 20 to about age 50, and then height gradually decreases.

Stop and Discuss

1. Identify a trend in your school.
2. What other trends are you aware of?

At Station 2 you found patterns between the sizes of different body parts. In humans the length of a person's forearm is about the same as the length of his or her foot. You could say that certain measurements "go with" certain other measurements. Patterns that involve connections like these are called **correlations.** Knowing correlations can help you make predictions. For example, if you know the circumference of your head, you can predict your height because there is a correlation between those two measurements.

There are other correlations besides those that involve body parts. For example, there is a correlation between the amount of exercise you get and how long you will live. In some areas of the country, birds fly south at the same time that the leaves on the trees change color. Because these two events occur at the same time, we say there is a correlation between them.

Stop and Discuss

3. Identify a correlation in your school.
4. What other correlations are you aware of?

At Station 3 you saw a pattern that probably seemed far removed from puzzles and body measurements. The swinging motion of the pendulums transferred back and forth—first one pendulum moved, and then the other pendulum started moving while the first one slowed down. This happened over and over. This type of repeating pattern is called a **cycle.** You have experienced other cycles in your life as well. For example, every day you can expect daylight and darkness. This pattern repeats day after day. In addition, unless you live near the equator, you can expect seasons to change every year. Even the years are a cycle, as you found in the investigation Driving Day. On our calendar a particular date (such as January 2) will shift forward 1 day per year until 7 years later, when January 2 will be on the same day again. As long as we keep the same type of calendar, this cycle will repeat over and over.

Stop and Discuss

5. Identify a cycle in your school.
6. What other cycles are you aware of?

Now you know more than just what a pattern is. You know about three kinds of patterns: trends, correlations, and cycles. You will have more opportunities to identify and use these types of patterns throughout the rest of the unit.

Stop and Discuss

7. How are trends, correlations, and cycles similar?
8. How are trends, correlations, and cycles different?

S I D E L I G H T

Footprints from the Past

In the 1950s and early 1960s, Louis and Mary Leakey, along with other scientists in Africa, were just beginning to unearth a puzzle dating from a time long before recorded history. They found some humanlike skeletons they were curious about. By the structure of the bones, they could tell that these humanlike animals were not apes. These animals walked upright as we do. But the skeletons were very old and not complete, so the scientists could not tell how tall these creatures had been.

But in 1978, scientists working with Mary Leakey revisited the site and made another discovery. In rock that had once been soft, damp, volcanic ash, they found trails of footprints from long ago that appeared to be made by humanlike creatures. First the scientists measured these footprints. Then they were able to estimate the height of these creatures by using a correlation that exists in humans between height and foot length. That is, the foot length of a person generally is about 15 percent of that person's height. The scientists concluded that one of the travelers was about 1.4 meters (about 4 feet, 6 inches) tall, and the other was about 1.2 meters (about 4 feet) tall.

Figure 1.15

(a) Cross section of a nautilus shell. (b) A waspcomb.
(c) Fallen maple leaves in the snow. (d) Car tire with ice.
(e) Sunflower.

CONNECTIONS:
Patterns in Nature

Many patterns exist in nature. Look at the following photographs and at the objects your teacher provides. In your notebook list as many patterns as you can identify in these photographs and objects. What type of pattern is each one?

(a)

(b)

(c)

(d)

(e)

Figure 1.16

This is a photograph of the Apollo 15 space command module *Endeavour* as seen from the Lunar Execution Module.

INVESTIGATION:
Moon Watch

Now that you know more about patterns of change, your class will be observing a very familiar pattern: the changing phases of the moon. In this investigation your challenge is to keep track of the pattern. In Chapter 4 you will be using the data you gather to explain the whole pattern.

Materials

For the entire class:
- 1 calendar
- small box with dated slips of paper

For each student:
- 1 Moon Watch Chart
- 1 Parent/Guardian Letter
- tape, glue, or stapler

Procedure

1. Take a copy of the Moon Watch Chart and a copy of the Parent/Guardian Letter.

   Notebook entry: Be sure to include today's date and the investigation title.

2. Attach your Moon Watch Chart into your notebook.

   Use tape, glue, or staples.

3. Fill out the first square of your chart, based on your teacher's moon watch report.

Working Environment

You will work in your cooperative teams of two. In addition to the role of Team Member, you will need a Communicator and a Tracker. Move your desks together or sit somewhere so that you face each other. Practice the social skill Move into your groups quickly and quietly.

4. Draw your moon watch assignment from the small box.

   Each Team Member should draw a date.

5. Record your date and your partner's date.

   Notebook entry: This also will go on the class calendar.

6. Fill in the blanks of your Parent/Guardian Letter.

   Remember to get your letter signed.

7. With your partner work out a strategy for remembering your dates. Record your strategy.

   Notebook entry: When it is your turn to do the moon watch report, have one partner draw on the chalkboard and one partner describe the moon. You will do two moon watch reports.

Wrap Up

Respond to the following questions.

1. Since the first time you practiced moving into your groups quickly and quietly, how much has your team improved: not at all, some, or a lot?

2. On your Moon Watch Chart, record the moon watch report that other students give each day.

INVESTIGATION: Making Music

What do foot measurements and music have in common? Both have patterns. In this investigation you will see how you can use patterns to make music. Try to identify the types of patterns you hear. If you cannot remember the differences among trends, correlations, and cycles, refer back to the reading Types of Patterns.

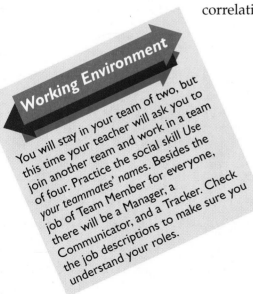

Working Environment

You will stay in your team of two, but this time your teacher will ask you to join another team and work in a team of four. Practice the social skill Use your teammates' names. Besides the job of Team Member for everyone, there will be a Manager, a Communicator, and a Tracker. Check the job descriptions to make sure you understand your roles.

Materials

For each team of four students:

- 2 rulers (15-cm) or metric tapes
- 4 glass bottles of the same size and shape (one per Team Member)
- 4 beakers or small cups filled with water
- 1 bowl (size of 1-qt margarine tub, 800 mL) (for waste water)
- 4 pieces of masking tape (each about 10 cm long)
- 8 cotton balls (two per Team Member)
- 1 small container of diluted bleach solution (about 20 mL)
- 4 pairs of goggles

Procedure

1. Write your teammates' first and last names in your notebook.
2. Pick up the materials.

   This is the Manager's job; the Tracker may help if needed.

   > ▲ **CAUTION:** Glass containers can break easily. If water spills occur, be careful not to slip. Clean up water immediately.

3. Distribute the materials.
4. Use a piece of masking tape to label your bottle with your name. Put on your goggles.

   Each Team Member should have a bottle, a cotton ball, and a piece of masking tape.

   > ▲ **CAUTION:** Hazardous substance. Wear eye protection. If the bleach solution splashes on your skin, wash it off immediately and tell your teacher.

5. Dip the edge of your cotton ball into the bleach solution and wipe the top 5 cm of your bottle to disinfect it.

   Work only on your bottle and throw away your cotton ball in the proper waste container when you are finished. Then you may take off your goggles, but remember to put them back on before you use the bleach solution in step 14.

6. Hold the empty bottle and blow over the top of it.

   Work quietly so that others can hear the sounds from their bottles.

7. Use your beaker or cup to add a small amount of water to your bottle.

   The water level should be approximately 1 cm deep.

8. Again hold the bottle and blow over the top of it. Describe any changes in what you hear.

   Notebook entry: Record your observations.

9. Determine how the pattern of sounds changes when you vary the amount of water in the bottle.

   Try to use small amounts of water at a time.

10. Next try gently tapping your bottle to generate a different pattern.

    Use your pen or pencil.

11. Describe the types of patterns you observed.

   Notebook entry: Record your observations.

12. Compare the sounds of all the bottles in your team.

   Only blow over your own bottle. Remember to use each other's names.

13. Use all of your team's bottles to play a tune.

   Try "Mary Had a Little Lamb." Listen politely to your teammates and express your ideas aloud as your team composes a tune.

14. Clean your bottle again, using another cotton ball dipped into the bleach solution. Remove your name label.

   Throw away your tape and cotton ball in the proper waste container. Return all materials.

Wrap Up

Write the words "Wrap Up" in your notebook. Discuss the following questions with your team and record your answer for each question in your notebook. Be sure each of you can explain your answers in a class discussion.

1. Did you notice any difference in the pattern when you tapped on your bottle, compared with when you blew over the top of it?

2. Describe the correlation between the height of water in the bottle and the sound you heard.

3. Describe any trends you observed.

4. How could you create a cycle with the bottles?

5. Were you able to remember and use your teammates' names?

6. Why is it helpful to remember and use each other's names when you are working in a group?

CONNECTIONS:
What Do You Know about Patterns?

The focus question for this unit is, How does my world change? Recognizing patterns of change can help you answer that question. In this chapter you have been exploring patterns of change. First you solved puzzles by discovering patterns. Then you learned about three kinds of patterns and how to recognize them. What is important for this chapter is that you can identify patterns. Part of that skill is seeing patterns on graphs. Another part is recognizing patterns in descriptions, events, or pictures. Use the information from this chapter to answer the following questions.

1. Look at the graph in Figure 1.17 People's Pattern of Eating Vegetables and answer the following questions.
   a. According to the graph, how has people's pattern of eating fresh vegetables changed?
   b. What kind of pattern is visible in the graph (a trend, cycle, or correlation)? Explain your answer.

2. Read the following and answer the questions. Isaac decided to go rowing every day. Each day he went out at 4:00 P.M. Each day he came back at 6:00 P.M. He repeated this activity each day for 3 months.
   a. What kind of pattern is this? Explain your answer.
   b. If Isaac exercised every day and grew stronger, what kind of pattern occurred? Explain your answer.

3. Describe a pattern that is a cycle.

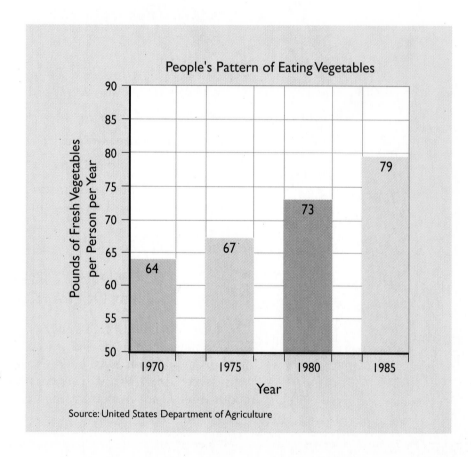

Figure 1.17

This graph shows how people's pattern of eating fresh vegetables changed between 1970 and 1985.

4. Use the following scale to rate your understanding of patterns:

  1 = I can't explain anything about patterns.

  2 = I can give an example of one kind of pattern.

  3 = I can define all three kinds of patterns.

  4 = I can explain all three patterns to myself.

  5 = I could teach others about all three patterns.

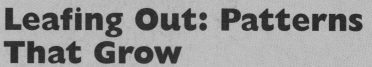

Leafing Out: Patterns That Grow

By now you probably realize that you have been seeing patterns all your life. Now that you are better at recognizing patterns, consider these questions: Do patterns always stay the same? If not, what makes them change? How can you find out what things affect patterns? In this chapter you will have a chance to answer these questions.

CONNECTIONS:
If I Were a Magnolia Tree

Move your desks side by side to work with your partner on this Think-Pair-Share activity.

Think: Imagine what it would be like to be a green, leafy plant such as a magnolia tree or a lilac bush. What do you think you would need in order to grow?

Pair: With your partner discuss your ideas about plant growth. Record all your ideas on one list in one of your notebooks.

Share: Contribute your ideas to a class discussion. You will create a large list on the board.

INVESTIGATION:
Beanstalk I

In the connections activity If I Were a Magnolia Tree, you and your classmates generated ideas about what plants need in order to grow. In this investigation you will begin testing your ideas by planting seeds and observing what happens.

Working Environment

You will work in your same cooperative teams of two. Besides Team Members, additional roles you need are Communicator/Tracker and Manager. Work at a table or push your desks together to form a table. Practice the social skill Move quickly and quietly into your groups.

Materials

For each team of two students:
- 3 bean seeds
- 1 paper cup
- potting soil
- water supply
- $\frac{1}{2}$ paper towel
- beaker or cup for water
- 1 metric ruler
- masking tape—a strip long enough to label your team's cup
- 1 small tray with gravel or one large tray with gravel for several teams
- 2 pencils

Procedure

1. Read through the list of materials and the procedure.
   Make your notebook entry.
2. Collect the materials.
3. Label the paper cup "C" for control. Put your initials and your teammate's initials on your cup.

4. Poke three holes into the bottom of the cup.

Use your pencil or a pen to do this.

5. Cover the bottom, inside of the cup with the wet, folded paper towel.

Use a half-sheet of a paper towel and moisten it with water.

6. Fill the cup ²/₃ full with potting soil.

Hold the cup over a tray. You can do this at the location your teacher designates.

7. Pour water into the cup until the soil is soaked.

Let any extra water drip onto the tray.

8. Measure 2 centimeters down the side of a pencil, beginning at the tip of the eraser. Mark this, using another pencil, to create a measuring stick.

You should now be back at your desk.

	Control plants (Beanstalk I)	Experimental plants (Beanstalk II)
Date planted		
Date sprouted		
Number of sprouts		
Number of days for sprouting to occur		
	Observations	
Day 1	No growth yet	
2		
3		
4		
5		
6		
7		
8		
9		
10		
11		
12		
13		
14		

(Add more days on as you need them.)

Figure 2.1

By using a chart like this one, you will be able to keep track of your plants' growth.

9. Poke three holes that are 2 centimeters deep in the soil.

   Use the measuring stick you created in step 8 for this.

10. Put one bean seed into each hole. Cover the seeds with soil.

11. Describe the contents of your control cup.

    Notebook entry: Record this description in your notebook. If you have followed the procedure, your control cup should have three bean seeds each planted 2 centimeters deep.

12. Decide on the best location for the control cups.

    Do this as a class.

13. Place your cup at that location.

14. Clean up materials as your teacher indicates.

    Practice moving quickly and quietly as you clean up.

Wrap Up

In your own notebook write today's date and the words "Wrap Up." Draw a chart like the one in Figure 2.1 and fill in Day 1 with "No growth yet." Now discuss the following as a team and record your answers.

1. If you are doing well moving quickly and quietly into your groups, write one thing you can do to improve further your use of this skill.

2. If you are not doing well moving quickly and quietly into your groups, list three things you could do to improve your use of this skill.

READING: What Affects Patterns?

At the beginning of this chapter, we asked you to imagine what a magnolia tree needs in order to grow. You might have thought about other trees you've seen and realized that all growing plants need many of the same things. Based on what you studied about patterns in Chapter 1, you might recognize that plant growth is a pattern.

For example, some trees show a cycle. Every spring they grow new leaves, and every autumn the leaves change color and fall. With the passing of the seasons and the beginning of another spring, the cycle continues.

If you look at the stump of a tree that has been cut down, you will see tree rings. These tree rings are also a pattern of growth that people can observe. Each ring represents 1 year of growth for every year of the tree's life (see Figure 2.2). Each year the tree grows outward, as well as upward, creating a new layer of living material surrounding the older layers. These layers appear as rings. Notice

Figure 2.2

In this photograph you can see the pattern of tree rings. Each year as the tree grows upward and outward, a new layer of living material is added around the older layers.

that some of the rings in Figure 2.2 are wider than others. This is because the same tree can grow at a different rate from one year to the next. In the years the tree grows quickly, the ring it forms is wider than the ring it forms when it grows slowly. Hence, there is a correlation between how quickly trees grow and how wide their rings are.

Not only can a certain tree grow at different rates, but different trees grow at different rates. For example, pine trees in Louisiana's warm, moist climate grow much faster than pine trees in Colorado's cold, dry climate. But this example involves trees in different locations. Even in the *same* location, different types of trees will grow at different rates. For example, spruce trees growing in the same forest as lodgepole pines grow more slowly than the lodgepole pines.

As you can see, many things influence the pattern of plant growth. The things that influence patterns are called **factors.** This reading describes several factors that influence the pattern of plant growth.

Stop and Discuss

1. Describe the factors you read about that affect the pattern of plant growth.
2. What additional factors might affect the pattern of plant growth?

Green Thumb Careers

Does your city have flower beds in parks and other places? A greenhouse technician designs, plants, and maintains these flowers and shrubs. A greenhouse is a glass-walled building that also has a glass roof to let in light. It is typically kept very warm. The greenhouse worker starts new plants by planting seeds, bulbs, and leaf or stem cuttings. If the plants seem to be doing well in the greenhouse, the technician might later transplant them outside. When designing flower beds, the greenhouse technician must select flowers and shrubs appropriate for the area. This takes knowledge of plant characteristics and requirements. The greenhouse technician also needs to know about soil types, plant disease, and insect problems and treatments. Greenhouse technicians have at least a high school diploma and some experience with plants.

Floral designers put together arrangements of flowers, leaves, and sometimes twigs. They usually work in florist shops or in the floral department of a supermarket. A floral designer must be familiar with the names of plants and with the characteristics and requirements of different plants. A career in floral design requires at least a high school diploma. Many community colleges offer courses in floral design.

Scientists who study plants are called botanists. Some botanists do research to develop new flower or vegetable varieties, bigger and better trees, or new methods of pest and disease control. This career requires a bachelor's degree or often a master's or doctorate degree in biology or botany.

READING:
Controlled Experiments

Sometimes it is difficult to single out what factors are influencing a pattern. But there is a way to determine whether or not a factor is affecting something.

For example, suppose you are raising fish in a pond. One day you decide to cut the amount of food you give them in half and move them to a different pond. If the fish stopped growing, which factor would you say was causing the change—the reduction in food or the change to a new pond? You might guess that it was the food, but it would be only a guess. To be certain, you would have to change only one thing at a time.

To see whether the change in food caused the fish to stop growing, you would have to cut the amount of the fish's food in half but leave them in the same pond. To see whether a new pond caused the fish to stop growing, you would have to move the fish to a new pond but keep feeding them the same amount of food.

Let's say that Rosalind was doing an experiment with fish. She wanted to find a factor that would affect their pattern of growth. She got two identical fish tanks and placed the same number of the same kind of fish in both tanks. She kept the water in both tanks at the same temperature and fed them all the same kind of food. The only factor Ros varied was the *amount of food* she was feeding them. When you change only one factor and leave all the other factors the same, the experiment is called a **controlled experiment.**

Rosalind had varied only the *amount of food* in her two tanks and therefore could observe whether or not it was the amount of food that affected the pattern of growth.

Marie and Isaac also were interested in factors that might affect the growth of fish. Marie wanted to know whether or not water temperature affected fish growth. Isaac wanted to know whether or not the size of the fish tank affected fish growth.

Marie also set up a controlled experiment using two fish tanks. She bought the same kind of fish, food, and tanks that Rosalind bought. She kept the conditions the same for both fish tanks except that she varied the *water temperature.* In one tank she kept the water at room temperature, and in the other tank she kept the water warmer.

Isaac also set up a controlled experiment using two tanks. He kept the number of fish, the type and the amount of food, and the

water temperature the same. The factor that Isaac decided to test was the *size of the tank*. He used one small tank and one large tank.

One day Al stopped by to look at what they were doing. He saw six fish tanks. Ros had two, Marie had two, and Isaac had two. They described their results to Al. Ros found that the fish who received more food were larger. Marie found that the fish in the cooler water were larger than the fish in the warmer water. And Isaac found that the fish in the larger tank had grown more than the fish in the smaller tank. Because Isaac, Marie, and Rosalind had done controlled experiments, their results meant something. When they each saw a difference in their fish, they were better able to explain what factors affected the pattern of fish growth.

As you may already know, fish can take a long time to grow. Some plants, however, can grow very quickly. In the next investigation you will conduct a controlled experiment to determine what factors affect plant growth.

INVESTIGATION:
Beanstalk II

In the last investigation you and your classmates used the same procedure for planting bean seeds. Every team used the same kind of seeds, cup, and soil. Then you all put your cups in the same place.

Now you and your partner will have a chance to do something different from other teams. Can you think of different factors to vary as you grow new plants? That is what you will try in this investigation.

Working Environment

Establish your environment as you did in Beanstalk I. Trade roles, but practice the same social skill Move into your groups quickly and quietly.

Materials

For each team of two students:
- 3 bean or 3 radish seeds
- 1 paper cup
- potting soil
- water supply
- ½ paper towel
- 1 metric ruler
- masking tape
- pencil

Procedure: Part A—The Social Skill

1. Compete with other teams to move into your group the most quickly and quietly.

2. When you are settled in your group, let your teacher know as quietly as you can that you are ready.

3. As a class have a brainstorming session to come up with a reward for the winning team.

Procedure: Part B—The Activity

1. With your partner discuss and choose what **factor** you will test. Make sure both of you have a chance to express your thoughts and ideas aloud.

   This factor should make the seeds you plant today grow differently from the seeds in your control cup. For ideas, look at the list on the board.

2. Decide what *one* change you can make in your second cup, your experimental cup, to test that factor.

   Notebook entry: Record the factor you chose and the change you will make.

3. Predict how this change will affect your plants.

   Notebook entry: Record your predictions.

4. Tell your teacher what factor you chose.

   This is the Communicator's role.

5. Plant the seeds in your experimental cup according to your decisions.

6. Label your experimental cup.

   You may label it with a name, letter, or number and also the Team Members' initials.

7. Put your experimental cup on a tray in the appropriate location.

   Your location might be different from that of your control cup.

8. Clean up the materials.

Wrap Up

Write the words "Wrap Up" in your notebook. Discuss the following questions with your partner and write your answer to each of the questions in your notebook. Make sure each Team Member can explain your answers in a class discussion.

1. Predict how many days your plants will take to sprout and which of your two cups will show more growth. Figure 2.3 shows bean plants at two different stages of growth.

2. Each day for the next week, look for any correlations you can find between your plants' rates of growth and the factor you are testing. For example, if one of your plants is getting more water, does getting more water correlate with the seeds sprouting sooner or the plants growing faster?

3. From your observations would you say that you and your partner are expressing thoughts and ideas better now than you were in the last investigation?

Figure 2.3

If you were able to see beneath the soil at different stages as your plants grow, you would see something like this.

 READING:
Cause and Effect in My Life

One morning as you ride the bus to school, an accident occurs on the highway. Traffic is delayed for 15 minutes. Fortunately no one is hurt. Unfortunately, as a result of the delay, you are late to your first class. Your teacher listens as you explain what happened.

Think about this situation for a moment. When one thing makes another thing happen, you say the first thing caused the second thing, the effect. You explained that you were late (the **effect**) because an accident delayed your bus (the **cause**). A cause is any event, circumstance, or condition that brings about a result or an effect. Because you understood the relationship between cause and effect, you could explain to your teacher why you were late.

Stop and Discuss

1. Give two examples of cause and effect in your life.

Scientists often investigate cause-and-effect relationships. What causes the common cold? What causes bean plants to grow at different rates? What causes the changes we observe in the shape of

Human Growth Hormone

We have looked at how patterns of growth may change in plants. But is it possible to change the pattern of growth of humans? The hormone called human growth hormone, or HGH, greatly affects the patterns of growth that humans show. We all need human growth hormone to grow to our own adult heights. Some people produce less of this hormone than other people, and these people are shorter. Some people produce a larger amount and they grow taller. In the past, when people's bodies produced too much human growth hormone, they were susceptible to serious diseases such as diabetes, and they often died young.

Today, human growth hormone can be manufactured artificially and then prescribed by doctors for people whose bodies do not produce enough HGH. The problem today, however, lies in the potential for the abuse of human growth hormone. Some athletes who want to grow bigger and stronger are willing to buy illegal supplies of HGH. However, if they inject too much of this hormone, they might be condemned to an early death because of changes the hormone might bring about in their bodies.

the moon? All of these questions seek explanations (causes) for patterns (effects) we see in nature.

Yet two things can be related and still not be a cause-and-effect relationship. Here is one example: In North America certain birds fly south at the same time of year that maple leaves change color. Does this mean that the birds cause the maple leaves to change color? Or do the changing leaves cause the birds to fly south? The answer to each question is no—neither event causes the other event to happen.

When two things are related but not necessarily cause and effect, we say they are **correlated.** The following are two examples of things that are correlated but not necessarily cause and effect. Read these examples and then answer the questions with your teammate.

Example 1

A local track team has been training vigorously for over a month. One student has a parent who works for Disgusto Cereal Company. The student brings a case of free samples to the track team. The next day all the students who ate Disgusto Cereal for breakfast had the fastest times in the track meet.

Stop and Discuss

2. Did Disgusto Cereal cause the fast times?

3. How could you explain the correlation between Disgusto Cereal and fast times at the track meet?

Elaborate

Example 2

A scientist studied a group of people to determine their patterns of watching television. She found a correlation between watching TV and health: People who watched TV were healthier than people who didn't. This correlation puzzled scientists. Could watching TV cause good health? Further investigation showed that many people in this study rode exercise bikes while watching TV.

Stop and Discuss

4. What might explain the correlation between watching TV and good health?

5. Is the relationship between watching TV and good health cause and effect?

6. What are some examples of things in your life that are correlated but are not cause and effect?

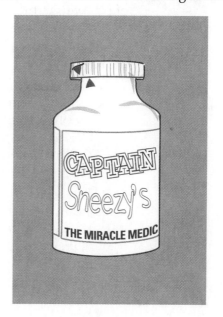

CONNECTIONS:
Captain Sneezy's Cold Cure

Work on this section in your cooperative teams of two. Each of you should work as a Team Member to accomplish the goal of this section.

Captain Sneezy's is a famous cold medicine. In fact the people who make Captain Sneezy's say that it will cure the common cold for all people. This type of amazing claim is common in advertising. Can you use what you know about controlled experiments to find out whether this amazing claim is true?

The following are experiments designed to find out whether the Captain Sneezy's claim is true. Take turns reading each of them aloud.

1. The first experiment, using two groups of people, tests both the regular and the extra-strength Captain Sneezy's. Half of the people are in Arizona and half are in Kansas. The group in Arizona gets the extra-strength medicine, and the group in Kansas gets the regular strength. The group in Arizona got better faster.

2. The second experiment uses people of all ages in the same location. The people are divided into two groupings: children and adults. Children are divided into two more groups. Half of them get a fake capsule that has no medicine. The other half of the children get Captain Sneezy's children's medicine. The adults all stay in one group. They all get regular Captain Sneezy's. The children got worse after taking Captain Sneezy's, and all the adults stayed the same.

3. The third study also tests regular and extra-strength Captain Sneezy's. The people are all in the same location. In this group some of the people are well and some are sick. Some of them get the regular medicine, and some get the extra-strength medicine. All of the people who got the regular dose felt good at the end of the experiment.

Discuss these questions with your partner and record your answers in your notebook:

1. Which experiments were controlled experiments?

2. Why do you think so?

3. Design your own controlled experiment to test the effectiveness of Captain Sneezy's. Be sure you and your partner are prepared to participate in a class discussion.

4. After the discussion as a class, list three factors that companies should consider when designing a controlled experiment of Captain Sneezy's medicine.

5. Look at the company's update graph in Figure 2.4.

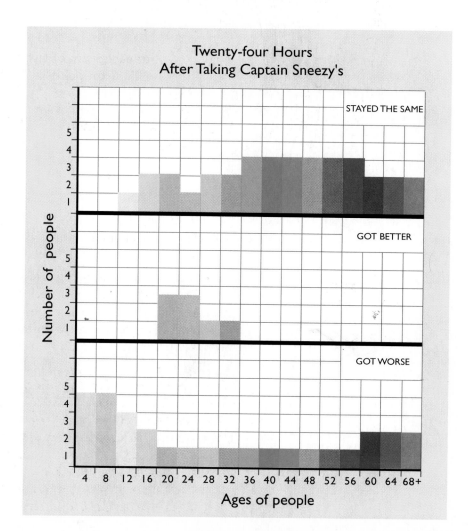

Figure 2.4

This graph shows what happened to people after they took Captain Sneezy's medicine. Compare the number and ages of the people who stayed the same with the number and ages of the people who got better and those who got worse.

Elaborate

6. Using the information in the graph, describe the changes that occurred. Answer these questions.

   a. What problems do you see in doing a controlled experiment on humans?

   b. Decide how you feel about whether or not scientists should use people in experiments. Now discuss how your feelings affect how you would design a controlled experiment.

INVESTIGATION:
Beanstalk III or Radishes on the Rise

By now you have made several observations of your team's plants. Recall that before you planted your bean or radish seeds you made a prediction about what would affect their growth. Then you designed an experiment to test your prediction. What observations have you made so far of the growth patterns in your plants? Do your observations match your prediction?

Working Environment

Set up the working environment as you did in Beanstalk I and Beanstalk II. Trade roles. The same person who was Communicator/Tracker in Beanstalk I should be the Communicator/Tracker again. The other person will be the Manager. You need the same kind of work space, and you will continue to improve your use of the skill Move into your groups quickly and quietly.

Materials

For each team of two students:
- team plants
- one 250-mL beaker or cup, filled with water
- 1 metric ruler

Procedure

1. Pick up your team's plants and bring them to your table or desks.

   Don't forget your notebook entry: Today's date and investigation title.

2. Observe whether or not any of your plants have sprouted yet.

   Notebook entry: Record or draw your observations.

3. Look for any differences between your control cup (from Beanstalk I) and your experimental cup (from Beanstalk II).

   Things to look for include:
   - *whether or not the seeds in the cups have sprouted,*
   - *how much the plants have grown, and*
   - *how healthy the plants look.*

4. Water your plants if they are scheduled for watering today.

5. Return the cups to their locations.

Wrap Up

Discuss the following questions with your partner and write your answer to each question in your notebook. Make sure each of you can explain your answers if your teacher asks you to.

1. Were your predictions accurate about how long it would take the plants to sprout?

2. If your plants have sprouted, which ones are bigger?

3. Do the predictions you made during Beanstalk II seem to be accurate?

4. Write a prediction about how much you think your plants will grow during the next week. (You might measure growth by counting the number of new leaves, for example.)

5. What are you doing that will allow your class to make accurate predictions about plant growth?

6. Do your predictions change when you have data from the entire class rather than just your team data? If so, how?

7. Rate yourselves on a scale from 1 to 10 (10 being the best) for your progress in practicing the unit social skill and the social skill for this activity. Compare your ratings with those of other teams and decide which teams have improved the most in practicing these skills.

 **CONNECTIONS:**
Plants All Over

Get into your teams of two. Discuss the following questions and be prepared to contribute to a class discussion.

1. Summarize the results of your plant growth experiments. Use a few sentences to describe the most important or interesting things that happened during the experiment.

2. How did the control plants and the experimental plants differ?

3. Knowing what you know now, what helpful hints could you make for other people who wanted to grow bean or radish plants?

4. Identify three patterns of plant growth. (You might need to listen to the class discussion first.)

Predictions: More Than Just a Guess

One morning when you leave for school, you wear a light jacket. You think it will stay sunny all day. At the end of the day, you find yourself walking home in the rain. How might you have avoided this annoying situation? Would patterns have been useful to you?

Patterns are useful for several reasons. Patterns help us solve puzzles. In Chapter 1 you solved puzzles such as a coded language and determined on which day of the week you would be eligible to drive. In Chapter 2 you identified factors that can change the patterns in plant growth or in how well a medicine works. In this chapter you will explore an additional way to use patterns. You will learn to use patterns to make predictions. Then you also might make fewer guesses in your life. You might even be able to predict better than the weather forecaster!

INVESTIGATION:
Finney's Funny Food

Hot chicken nuggets, crispy french fries, ice cold soft drinks, juicy hamburgers, and creamy shakes! Some of these foods might be your favorites, while others don't interest you in the least. If you like the food at a particular restaurant, you probably will return. You may establish a pattern of eating at the same restaurant several times a month. To stay in business, restaurant owners depend on their customers to return. In this investigation you will have the opportunity to look at a pattern and to predict what lies ahead for Finney's Funny Food.

Working Environment

You will work in cooperative teams of two. Push your desks together or sit side by side at a table. The roles you will use are Communicator, Tracker, and Team Members. Work on the social skill Speak softly so only your teammates can hear you.

Materials

For each student:
- 1 copy of Sales Results for the Second Six Months at Finney's

Procedure: Part A—The Social Skill

1. Move into your groups.
2. Construct in your notebook a T-chart labeled "Speak softly so only your teammate can hear you."

   It should have a "Sounds like" column and "Looks like" column.

3. As you discuss the social skill, record in your chart your team's ideas about how this skill would look and sound when you use it properly.
4. As you work through Part B of this investigation, try to be the most soft-spoken group.

Procedure: Part B—The Activity

1. Look at the pictures of the types of food served at Finney's.

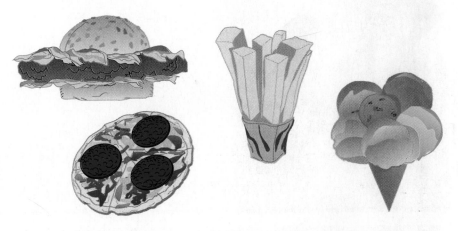

2. Read the following paragraph:

From the beginning Finney's was known for its big food. Finney's Food offers the largest food in town: Mammoth French Fries, Colossal Hamburgers, Gargantuan Ice Cream Cones, and Cosmic Pepperoni Pizza Topping. People flocked to Finney's for the first 6 months after its grand opening. Being a smart business person, the manager of Finney's decided to keep track of the sales data. She organized these data into several graphs.

3. Study Figure 3.1, Sales Results for the First Six Months at Finney's.

The graphs show sales data for January through June.

Figure 3.1

These graphs show the sales results of certain foods at Finney's from January through June.

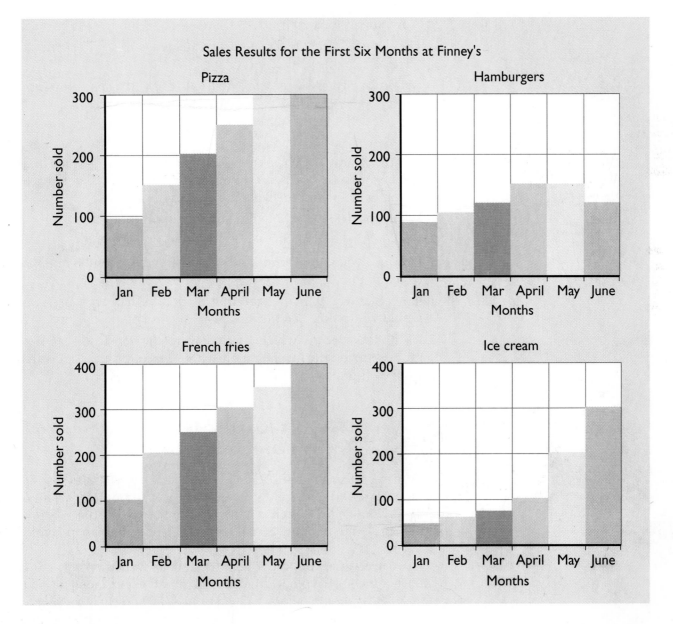

4. Discuss the patterns of sales results with your partner.

   Notebook entry: Write a comment about the pattern on each graph. Be sure to mention whether you see a trend, a cycle, or a correlation.

5. Predict what patterns the graphs might show for the next 6 months.

   These graphs will show sales for July through December. Think about these things: What items will sell well in cold weather, what items will sell well in hot weather, what items sold well before.

   Notebook entry: Record your predictions. You could draw a graph or write a sentence.

6. Participate in the class discussion.

   Be prepared to present your predictions to the rest of the class.

7. Study the graphs your teacher hands out.

   These graphs show the actual sales results of the second 6 months at Finney's.

   Notebook entry: Write a comment about the pattern you see in each graph.

Wrap Up

Discuss the following wrap-up questions with your partner and write your answer to each question in your notebook. Be sure each of you can explain your answers in a class discussion.

1. What was the best-selling food at Finney's during June?

2. Explain how you made your predictions of future sales.

3. How did your predictions for sales during the second 6 months compare to actual sales?

4. Now that you have 12 months of data, what type of sales would you predict for the next year?

5. As a class decide which groups were the most soft-spoken groups during the investigation. Choose an appropriate reward for these groups.

INVESTIGATION:
The Power of Attraction

Sometimes patterns are surprising. Think back to the sales results from Finney's. Did you accurately predict the sales for the second 6 months? Knowledge about a pattern can help you make better predictions. And making better predictions can help you avoid surprises.

How much do you know about magnets? Whether you have pulled magnets off your refrigerator door or experimented with magnets in school, you probably know that some magnets are stronger than others. But is there a pattern to which ones are stronger? In this investigation you will have the opportunity to test your knowledge about magnets by making some predictions. See whether you encounter any surprises!

Materials

For each team of two students:

- 1 small-sized magnet
- 1 medium-sized magnet
- 1 large-sized magnet
- 1 box of small metal paper clips
- transparent tape (four 1-cm strips)

Procedure

1. Draw a chart in your notebook like the one in Figure 3.2.

   Notebook entry: Be sure your entry also includes the title of the investigation and today's date.

2. Pick up all the materials, except the large magnet.

3. Hold the small magnet and add seven paper clips, one at a time. Then stop adding paper clips.

   Have the Manager hold the magnet as shown in Figure 3.3 while the Communicator adds the paper clips.

Working Environment

You will work cooperatively in your same teams of two. The roles you need are Team Members, Manager, and Communicator. Work with your desks pushed together or sit side by side at a table. Continue to practice the social skill Speak softly so only your teammate can hear you.

Figure 3.2

This chart will help you record your predictions and the results during this investigation. Copy this chart into your notebook.

PREDICTION CHART

Magnet size	Prediction (# of paper clips I think the magnet will hold)	Results (# of paper clips the magnet actually held)
Small		
Medium		
Large		

Figure 3.3

The Manager should hold the magnet like this, between the thumb and index finger.

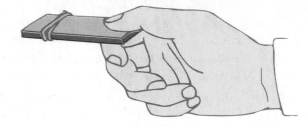

Explore

4. Predict the total number of paper clips you think the small magnet will hold.

   Notebook entry: Record this prediction in your chart.

5. Continue adding and counting the paper clips until some paper clips begin to slip off.

   You may put them on in any way, as long as they are below the band marked on the magnet. Stop when one paper clip has slipped off at least three times.

6. Record the total number of paper clips the small magnet held.

   Notebook entry: Write this in your chart.

7. Pull the paper clips off the small magnet and set them aside.

   Do not use these clips again.

8. Predict how many paper clips you think the medium-sized magnet will hold.

   Notebook entry: Record this prediction in your chart.

9. Hold the medium-sized magnet just as you held the small magnet. Repeat steps 3 through 7, using the medium-sized magnet.

   Remember to use new paper clips and to add them only below the band marked on the magnet.

10. Ask your teacher for the large-sized magnet.

    This is the Manager's job. Do not test the magnet yet.

11. Use the pattern you have seen to predict how many paper clips the large magnet will hold.

    Notebook entry: Record this prediction in your chart.

    STOP: Are you remembering to practice the unit social skill, Express your thoughts and ideas aloud and listen politely to others?

12. Begin adding paper clips to the large magnet.

    Use new paper clips and hold the magnet as before. If you want to change your prediction during the investigation, enter your second prediction in the appropriate place on your chart.

13. Record the final number of paper clips the large magnet held.

    Notebook entry: Record this in your chart.

14. Clean up the materials.

    Keep the paper clips you used separate from the paper clips you did not use.

Wrap Up

Discuss the following questions with your partner and write your answers in your notebook. Be sure to incorporate your partner's ideas into your answers.

1. What information did you use to make your predictions?

2. How accurate was your prediction for the medium-sized magnet?

3. Did the strength of the large magnet surprise you? Why?

4. If you revised a prediction, explain what caused you to change your mind.

5. Describe any patterns you observed while using the magnets. Be sure to identify any trends, cycles, or correlations.

6. Use your own rating system to rate your team on how well you are practicing the social skill, Speak softly so only your teammate can hear you.

7. What specific things can you do to improve your rating?

 READING:
Making Accurate Predictions

So far in this chapter, you have used patterns to try to predict fast-food sales and how many paper clips a magnet would hold. Now think about the accuracy of your predictions. What happened to Finney's sales, compared to your prediction? What happened with the magnets?

Sometimes a pattern seems very clear and the outcome seems obvious. But the expected result does not always happen! When unexpected results occur, the people making the predictions probably needed more information. To make an accurate prediction, you need a sufficient **quantity** of information. People cannot make accurate predictions without enough information. To be sure that their information is useful, people also must make careful observations and take accurate measurements. This provides information of good **quality**. Predictions require information of both sufficient quantity and good quality.

After testing the small- and medium-sized magnets, you attempted to predict how many paper clips the large-sized magnet would hold. It's possible, however, that you based your prediction on too little information. After trying the first two magnets, maybe you thought you saw a pattern: the larger the magnet, the stronger its power of attraction. If you had been able to test several large magnets before making your prediction, you might have discovered that large magnets can be very weak. In other words, you had good information—you just didn't have *enough* information to make an accurate prediction.

Now reconsider your predictions about sales at Finney's Foods. You might have predicted that sales would continue to increase because the graphs indicated a steady upward trend in total sales. By gathering *more* information on total sales, you might have revised your predictions for the following year. Yet you still don't know from total sales whether customers liked Finney's Foods and

Sidelight: Lodestone

Have you ever tried to find your way in a strange place? Without landmarks or signs, or particularly if you are traveling after dark, it can be very difficult to know which way to go. You might become completely lost. But be assured that you are not the first person to struggle with this problem.

For hundreds of years people puzzled over the question of how to determine their location without the help of landmarks. They didn't know that a simple tool could have helped them. That tool was a magnet.

Since at least the seventh century B.C., people understood that magnets attracted metal objects. We have records of an ancient Greek legend that describes a magnetic island. According to the legend, if a ship sailed too close, the force of the island would pull out the ship's nails. But it wasn't until much later that people learned how to make use of magnets.

Using a magnet as a tool for navigation was first recorded by the Chinese in the 12th century A.D. They constructed a **compass.** This is an instrument that shows which direction a person is facing—north, south, east, or west.

Early compasses were very simple. A person would place a magnetic stone (a piece of magnetic iron ore) into a pan of liquid and watch to see which direction the stone turned. One end of the magnet would always point north. With this information a person could then determine which way was south, east, and west. By doing this, sailors could navigate without the help of the stars as guides. Modern compasses are not set in

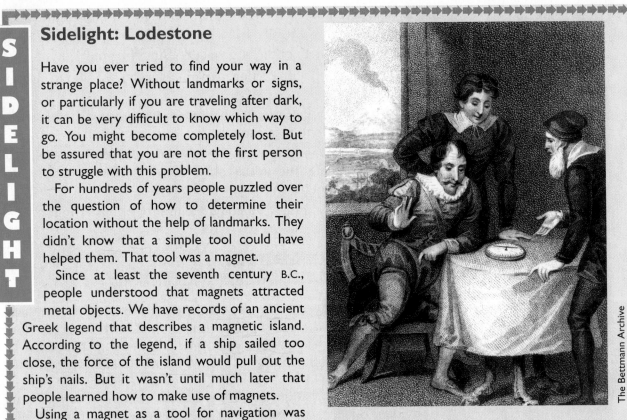

Although the Chinese were using magnets as a tool for navigation as early as the 12th century A.D., news of the compass did not reach England until the 17th century.

a pan of liquid, but they make use of the same phenomenon to show us which way is north.

When you align the point of the needle with the label "north" on the compass, you know

(Continued on next page.)

whether they will keep returning year after year. In this case you could make better predictions if you had a different type of information. Sometimes when you have several types of information to choose from, one type provides more useful information. When people have the types of information they need, we say that they have information of better quality.

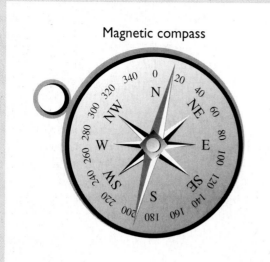

Magnetic compass

The needle of a compass is really a bar magnet. The needle is attached to the middle of the circle but turns freely so that the needle always points north.

which direction is north. Then, by reading the other labels on the compass you can tell which direction is south, east, and west.

By the 17th century, news of the compass had reached England. The English named the magnetic stones *lodestones,* meaning "leading stones." English scientists began to study these magnetic stones to determine how they worked. One scientist, William Gilbert, thought that the earth itself was a giant magnet and that objects such as lodestones became oriented in a particular direction because of the influence of the earth.

Today we know that many different materials can be magnetized, but the strongest magnets usually have iron in them. Lodestones, for example, contain the mineral magnetite, which has iron in it. Iron is a unique metal because of the way the particles in it can line up. When the particles line up in a certain way, this creates magnetism.

Scientists now understand that the earth does act as a giant magnet just as Gilbert thought. If you place a tiny magnet, a "tester," near a large magnet, the tester is attracted to (or repelled by) the large magnet. If you place this tester at different locations near the large magnet, it will be strongly attracted to (or repelled by) the large magnet. The entire region in which the large magnet influences the tester is called the field of the large magnet. In the example of the compass, the earth plays the role of the large magnet whose magnetic field influences the compass.

You can map magnetic fields, using a small magnet or a compass as a tester. Not only will the tester be attracted to or repelled by the magnetic field, but if you place the tester at a number of different locations near the magnet, you can map the strength of the attraction or repulsion, as well as the direction of the tester's orientation. You can use this map to predict the behavior of other magnetic materials placed near the magnet.

Humans cannot detect magnetic fields without the aid of tools such as the compass. But some other organisms are capable of detecting magnetic fields. For example, pigeons contain an internal compass that they use for long-distance navigation. And certain bacteria have internal compasses that appear to help them find a suitable living environment.

Stop and Discuss

1. Describe a time when you made a prediction based on too little information.

2. How would more information have changed your prediction you mentioned in question 1?

Figure 3.4

Businesses fail for many reasons. Sometimes the reason is that the owners don't have the quantity or the quality of information they need in order to make the best business decisions.

3. Imagine another fast-food restaurant like the one in Figure 3.4. Its owners had good information about sales for the first year. What other information could they have used to help them predict future sales accurately and perhaps have avoided going out of business?

4. If they had wanted responses from customers, what questions might they have asked?

We can try to make predictions about many things in our lives: sports events, grades in school, the success of a restaurant, and even the weather. You might take predictions of the weather for granted. Turn on the news and there is the forecast.

More accurate weather predictions are now possible because we have satellites that orbit the earth and send back photographs of clouds and weather patterns. This technology has improved both the quality and quantity of data that help scientists predict the weather.

Hurricanes always have been difficult for scientists to predict. Hurricanes can kill people and cause extensive property damage due to strong winds and flooding. People have known about such storms for thousands of years. But predicting them accurately has been possible only recently.

In 1938 the United States Weather Service monitored hurricanes and tropical storms by using reports from ships at sea. In mid-September of 1938, the weather service received word about a tropical storm off the coast of Florida. This storm easily might have become a hurricane, so they broadcast warnings, and people in Florida prepared for the storm. The ships monitoring the storm then came back to port. But no storm hit the coast of Florida. Weather forecasters thought that the storm probably had turned back out to sea where it would eventually die out.

The storm, however, had not died out. Unknown to everyone, the storm was moving directly northward. Instead of curving east and back out to sea, the storm bore down on New England. The wave of floodwater was so high (40 feet) that people first thought it was fog. Then they literally had to swim for their lives. Within a few hours after the storm hit land, it had killed 600 people and destroyed 60,000 homes.

Figure 3.5

Photograph shows the force of Hurricane Dora, 9 September 1964. Hurricanes have been difficult for scientists to predict. New technological developments, however, provide scientists with better quality and quantity of information. Now predictions are much more accurate.

The Bettmann Archive

Explain

5. How might the scientists have improved the accuracy of their prediction?

6. What might have been the result of a more accurate prediction?

Another tragedy occurred during World War II because of an inaccurate prediction. The Island of Malta, which lies in the Mediterranean Sea, was important to the British and their allies. The British had to find a way to defend the island. They dispatched an aircraft carrier with 14 planes aboard to Malta. Due to enemy air patrols, the aircraft carrier could not get very close to the island, so the pilots had to fly the rest of the way there. The planes, called Hurricanes, had a listed flying range of 521 miles, so when the aircraft carrier reached a point about 420 miles from Malta, the planes took off. But no one had calculated for a change in wind direction. Flying into the wind, the planes' range was even *less* than 420 miles. Miraculously five planes did reach the island, but the other nine crashed into the sea.

Stop and Discuss

7. How do the examples above show the importance of accurate predictions?

8. What are the characteristics of accurate predictions?

Figure 3.6

This is a photograph of a F6F Hellcat from World War II. In order to make accurate predictions about the flying range of a plane, you would need to have information about the wind conditions, as well as about the plane.

The Bettmann Archive

Explain

INVESTIGATION:
Will It Sink or Float?

So far in this chapter, you have used patterns to make predictions. In addition you have learned that you can make your predictions more accurate if you have enough information of good quality. In this investigation your challenge is to observe what happens with objects in liquids. Then use the patterns you see to make predictions.

Materials

For each team of two students:
- 25 mL Solution A
- 25 mL Solution B
- 25 mL Solution C
- 3 containers (30-mL clear cups, one for each solution)
- 1 piece of wood
- 1 piece of cork
- 1 piece of wax
- 2 pairs of goggles
- paper towels
- 3 pieces of masking tape

Procedure

1. Pick up the materials.

   This is the Manager's job.

2. Label each container.

   This will help you keep track of each solution. If it is already marked, you do not need to label it again.

3. Draw a prediction chart in your notebook.

   See Figure 3.7.

4. Fill in the first and third columns of the chart.

   In the first column record the type of material of each object. In the third column record a prediction about whether each object will sink or float in Solution A.

> ▲ **CAUTION:** Be sure to wear your eye protection.

5. Test your predictions for Solution A.

   Gently place each object into Solution A.

6. Record the results of each test. Then dry off each object.

   Notebook entry: Write "sinks" or "floats" in the Results column.

7. Predict which objects will sink and which ones will float in Solution B.

   Notebook entry: Record your prediction in your chart.

Working Environment

You will work in the same cooperative teams of two. Use the following jobs: Team Members, Manager, and Communicator/Tracker. Work at a table or form a table with your desks. Practice the social skill Move into your groups quickly and quietly.

PREDICTION CHART FOR SINK OR FLOAT

Object	Solution	Prediction	Results
1.	A		
2.	A		
3.	A		
4.	A		
1.	B		
2.	B		
3.	B		
4.	B		
1.	C		
2.	C		
3.	C		
4.	C		

Figure 3.7

Use a chart like this one for recording your predictions and the results.

8. Test your predictions.

   Make sure the objects are dry before and after testing.

   Notebook entry: Record your results.

9. Predict which objects will sink and which ones will float in Solution C.

   You might want to look at the patterns of results for Solutions A and B before making your predictions.

10. Test your predictions.

    Remember to dry each object before and after dropping it into the solution. Notebook entry: Record your results.

11. Return all materials.

    Recycle the solutions as directed.

Wrap Up

Write the words "Wrap Up" in your notebook. Discuss the following questions with your partner and write your answer to each question in your notebook. Be sure both you and your partner can explain the answers during a class discussion.

1. Describe any patterns you observed during this activity. (Try to be the team that comes up with the most patterns.)

2. How accurate were your predictions for Solution A?

3. How accurate were your predictions for Solution B?

4. How accurate were your predictions for Solution C?

5. What could you do to improve the accuracy of your predictions?

6. Describe a trend you saw in this investigation. You may choose from any of the three solutions.

7. Describe something about this investigation that surprised you.

8. Recall the social skill, Move into your groups quickly and quietly. Even though we did not ask you to work on that skill in this investigation, you should always use it. Look back in your notebook and find your answers to the Wrap Up for Beanstalk III or Radishes on the Rise. How would you rate your team now for using the skill, Move into your groups quickly and quietly: better, the same, or worse?

CONNECTIONS:
Minding Your Ps (Predictions) and Qs (Quality and Quantity)

Work individually during this connections activity.

In the investigations in Chapter 3, you made predictions based on the information available to you. You might have found that some of your predictions were more accurate than others. What made the difference? When your predictions were inaccurate, was it because you did not have enough information or because you did not have the type of information you needed?

Read the questions below and then review Finney's Funny Food, The Power of Attraction, and Will It Sink or Float? Look at the type of information and the amount of information you had in each investigation. Write the answers to the following questions in your notebook and be prepared to discuss your answers with your classmates.

1. Which investigations do you think had quality information but not a sufficient quantity of information?

2. Which investigations do you think had a sufficient quantity of information but not the appropriate quality of information?

3. Which of the investigations do you think had both sufficient quantity and quality of information?

4. Do you think any of the investigations had neither?

5. Rank these activities according to which had the most sufficient quantity and quality of information and which had the least sufficient quantity and quality of information. Write the rankings and your reasons in your notebook.

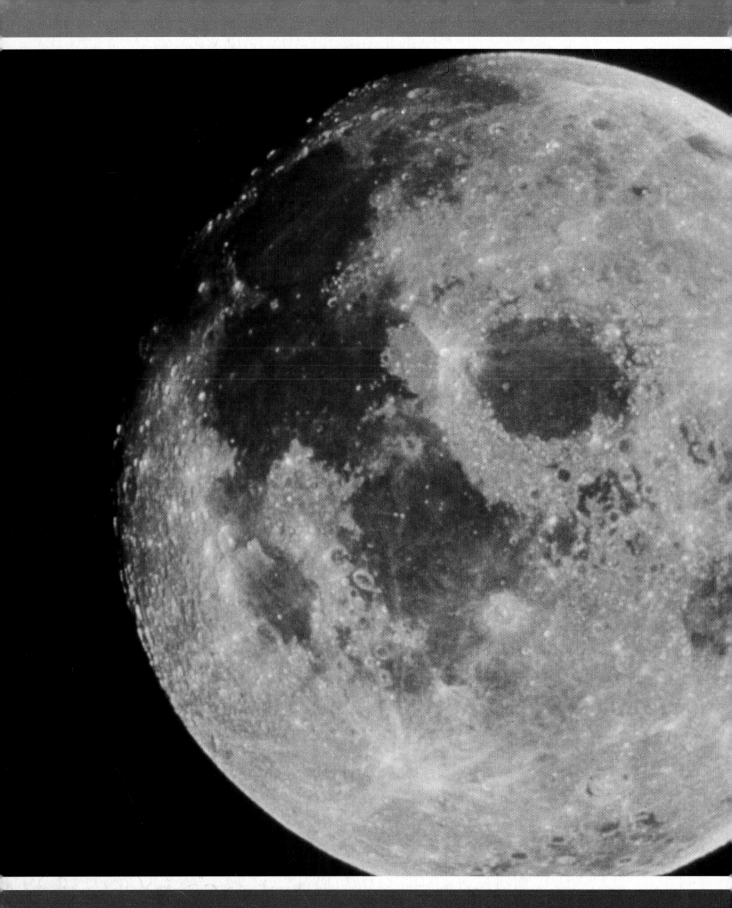

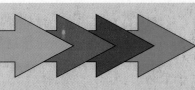

The Moon and Scientific Explanations

People have been curious about the cycle of the moon's apparent shape for thousands of years. People in many different lands often gathered in the evenings to tell stories. On some evenings they told their tales by the light of the moon. And when the moon mysteriously grew thin and disappeared, they told their stories to fill the darkness. They asked the same questions you might ask today: How did the moon get there? How does it move across the sky? Where does it go when it disappears?

READING:
Moon Legends—Another Way of Explaining Patterns

The stories people told long ago were important to them in ways that we may find difficult to understand. The stories were so important that they were told over and over again, and they were so powerful that the stories survived from one generation to the next.

The people who told the stories were trying to explain things they observed in the world around them. They felt the direct power of the sun and the moon in their lives. Storytellers in some cultures gave the moon a personality; it was attractive, stubborn, and changeable. Other cultures gave the moon a frightening, angry, and

The moon is like a fire in a hogon. When the door opens the fire glows and it is a full moon. When the moon is getting new, the blanket doorway is shutting.

Harley D. Ruiz
Tso Ho Tso Middle School
Ft. Defiance AZ

jealous personality. Still other cultures made the moon hopeful and happy. Sometimes the stories portrayed the moon as having a superhuman power and mystery. Regardless of the moon's personality, the stories always depicted the moon as having an important place in the lives of men, women, and children.

Many stories explained why the whole face of the moon did not shine all the time. An old African folktale tells of the keeper of the "great shining stone" (the moon). In this tale some special people called sky people are in charge of bringing food to the keeper of the great shining stone. Sometimes the sky people get tired of bringing food, however. When this happens, the keeper gradually covers the box that contains the stone. When fresh supplies of food come again, he gradually opens the box so that the stone can send its light down to earth.

Once long ago somewhere in the sky a person sleeps in the daytime and at night he wakes up and gets his weapons. His weapons are a bow and arrow. His bow is really bright. That makes the cresent of the moon. Sometimes he sleeps for a long time and that's a new moon. The arrow is the stars. When he shoots it, the stars come out.

Patrick Begay
Marlene Scott
Tso Ho Tso Middle School
Ft. Defiance AZ

CONNECTIONS:
Tell It Your Way

In Chapter 1 you looked at photographs showing patterns in nature. You probably also saw patterns in things your teacher brought to class. And you know of other patterns in our world too. You might have seen birds fly south before winter or noticed a rainbow after a storm.

Now it is your turn to make up a legend to explain a pattern that interests you. You may use any pattern you've seen or read about. If you can't think of a pattern to write about, your teacher will offer some suggestions. Before you write your legend, create a list of ideas, an outline, or a concept map of your story in your notebook.

As you write your legend, be imaginative but make sure you explain everything you observed or read about the pattern you chose. When you are finished writing, create a picture to go along with your legend.

Take turns sharing your legends with each other in the manner your teacher describes.

INVESTIGATION:
Moon Movies

Ancient people observed patterns in the changing shape of the moon. They revered the moon and followed its changes closely. In this investigation can you identify the pattern of change that made the moon so mysterious to early observers?

Materials

For each team of four students:
- moon charts from the moon watch
- 4 pairs of scissors
- 2 glue sticks or bottles of school glue
- 32 index cards, 3-by-5-in.
- 4 Moon Movie Flip Pages
- 1 metric ruler

For the entire class:
- at least one stapler

Procedure

1. In your notebook write the first and last names of all three of your teammates, today's date, and the investigation title.

   Notebook entry.

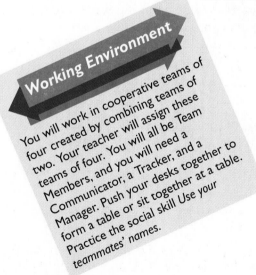

Working Environment

You will work in cooperative teams of four created by combining teams of two. Your teacher will assign these teams of four. You will all be Team Members, and you will need a Communicator, a Tracker, and a Manager. Push your desks together to form a table or sit together at a table. Practice the social skill Use your teammates' names.

2. Look at the moon watch charts.

   You drew these during the moon watch reports. The charts should be in your notebook.

3. Using each other's names, discuss with your teammates how the moon changes from a new moon to a full moon.

   A new moon is a moon you can't see, and a full moon is a fully lit moon.

   Notebook entry: Write several distinct changes the moon exhibits during this time. Make sure you have identified three key stages that fall between the new moon and the full moon.

4. Fill in five pictures of the moon, including a new moon, three key stages, and a full moon.

   Do this on your own Moon Movie Flip Page.

5. Decide on three key stages of the moon that occur between the full moon back to the new moon.

   Do this as a team.

   Notebook entry: Record your key stages.

   STOP: Review your unit skill.

6. On your Moon Movie Flip Pages draw those three key stages that occur between the full moon and a new moon.

   Again work individually on your Moon Movie Flip Page.

7. Cut out all the moon pictures.

   These are the pictures you drew in steps 4 and 6.

8. Glue each picture of the moon in the center of an index card.

9. Each pair should combine their cards, and sort them so that the changes in the moon are in order, beginning with a new moon and ending just before a new moon.

   You can have multiple cards for the same phase. Your group of four should have two moon movies.

10. Check each stack to make sure everyone agrees with the order of both stacks of cards.

    Each Team Member should take a turn doing this.

11. Staple the cards in each stack together.

    Measure the thickness of your stack of cards and record the thickness in your notebook.

12. Flip through the index cards, using your thumb and forefinger, and watch the moon movie.

    Each Team Member should take a turn doing this. While you are waiting for your turn, read the following Background Information.

Crescent...Crescent...Just the mention of it makes me hungry...Crescent roll...Crescent roll.

Background Information

Each of the different pictures that you drew in this investigation was a picture of a **phase** of the moon. The word phase comes from the Greek word *phaino,* which means "to appear or to bring to light." As the phase of the moon changes, more (or less) of the moon is "brought to light." If you start with a dark moon, the phase is called the new moon.

The next phase is a thin sliver of light in the shape of a crescent. This moon is called a crescent moon. When the moon appears as a shining half-circle, this phase is called the first quarter moon.

The phase when the bright portion of the moon is larger than a half-circle is called the gibbous moon. Finally, when a full circle of the moon is bright, the phase is known as the full moon.

There appear to be two trends to the changing shape of the moon: One trend is a change toward a full moon, and the other is a change back toward a new moon. The moon completes a full cycle from new moon to full moon and back to new moon every 29.5 days, or roughly every month.

If you don't watch the changes in the shape of the moon long enough, you might think that the pattern of change is a trend. By watching the phases for more than 1 month, you would be able to see the full cycle of change.

Wrap Up

Discuss the following with your team and write your answers in your notebook. Be sure each of you can explain your answers in a class discussion.

1. Describe what you saw when watching the moon movie.
2. Describe the pattern of change that occurs in the moon's shape.
3. Why is the moon's pattern of change called a cycle?
4. Record specific examples of how your team practiced the unit social skill.

 CONNECTIONS:
Moon Movies and Predictions

Work in your cooperative team of two in this connections section. Be sure both of you can explain your answers in a class discussion.

1. Imagine a whole year of moon movies. Approximately how many times would the cycle repeat?
2. Look back in your notebook and see how thick your moon movie stack was. How thick would your moon movie stack have been if you had made a movie that covered an entire year?
3. Explain why your information on the phases of the moon is of good quality and sufficient quantity.
4. What predictions can you make about the phases of the moon next year?

READING:
What Makes an Explanation Scientific?

Why do people invent legends and stories? If you think about the legends you read in this chapter, you probably will realize that the legends explained a pattern of change. As you told your legend, you were providing explanations for a pattern as well. Perhaps your story explained why we have day and night or why water moves downhill. In order to tell such a story, you must have observed a pattern and then developed an explanation for it.

Many explanations today no longer sound like legends or stories. In these explanations the moon is not a person and no white horses pull the moon across the sky. Today we explain many of the patterns in nature by gathering information and making predictions. Such explanations are called **scientific explanations.** They explain the cause of a pattern, and people base these explanations on information—that is, something they can hear, see, or experience.

Then people use this information to make a prediction. If they have sufficient quality and quantity of data, people often can make an accurate prediction about a pattern. This prediction helps them develop an explanation for that pattern. In Chapter 3 you learned how to make accurate predictions using data of good quality and sufficient quantity. Now you will see how you can strengthen your predictions using scientific explanations.

Stop and Discuss

1. What are the characteristics of a scientific explanation?
2. How is a scientific explanation different from a story or legend?
3. In which of the investigations that you have completed in this unit did you develop a scientific explanation?
4. Evaluate the following story. How is it like a legend? How is it like a scientific explanation?

   A mischievous, invisible elf put a spell on a dairy cow, causing her milk to taste bitter. All of her milk was made into ice cream and delivered to Finney's. No one bought much of that ice cream.

INVESTIGATION:
Explaining Phases

The moon is a familiar part of your surroundings. Even so, many people don't know how or why the appearance of the moon changes from day to day. In this investigation you will gather evidence that will help explain the relationships among the earth, moon, and sun. You will use this evidence to explain the patterns of change in the moon's appearance.

Working Environment

You will be working as an entire class. The skill to practice in this investigation is Express your thoughts and ideas aloud and listen politely to others.

Materials

For the entire class:

◼ 2 floodlights or light bulbs in fixtures

For each student:

◼ 1 Styrofoam™ ball, with an X marked on it, mounted on pencil

Procedure

1. Read through this procedure.

2. Stand in your assigned place.

   The teacher will tell you where to stand. Take your Styrofoam™ ball with you.

3. Hold your Styrofoam™ ball out away from you, above your head, with the X facing you so that you can see it.

   The Styrofoam™ ball represents the moon, and you represent the earth.

4. Stay in one spot and slowly rotate yourself one-quarter of a turn to the right. Then look at the ball.

   See Figure 4.1. Hold the ball at arm's length, higher than your head. As you turn and look at the ball, notice how the light falls on it.

Figure 4.1

Hold the ball with the X facing you. The ball represents the moon, and you represent the earth. The light represents the sun. Make one-quarter of a turn as you stand in one spot. Then stop and look at the ball. Notice what phase of the moon the ball resembles.

5. Rotate another quarter of a turn and look at the ball again.

   Again compare the ball to the moon.

6. Repeat step 5 two more times.

   You will have turned completely around once, while standing in the same place.

7. Make another full turn while holding the ball out but this time make only one-eighth of a turn at a time.

   You will be turning only half as far each time as you did before. You will make eight little turns to make one complete turn. After each little turn, look at the ball and compare it to the moon.

8. Turn in a circle again and demonstrate:

   - a full moon
   - a gibbous moon
   - a quarter moon
   - a crescent moon
   - a new moon

   Be certain that you know how to demonstrate each phase of the moon. If you do not, ask questions!

9. When you are finished, put away your materials.

Wrap Up

Discuss the following questions and write your answers in your notebook.

1. What approximate length of time does it take for the moon to complete all the phases? (Think back to the moon watch you have been doing.)

2. Explain why we see different phases of the moon from the earth.

3. What do you think is the source of the moonlight you see?

4. Look at the diagrams your teacher shows you on the overhead transparency. Describe how the phases of the moon are different from before.

5. Explain how these different shapes of the moon phases might occur.

6. Come up with a class rating for how well you practiced the social skill, Express your thoughts and ideas aloud and listen politely to others. Use a scale from 1 through 10. Discuss how you could improve your rating.

So, THAT explains why a half-circle is called a quarter.

Elaborate

CONNECTIONS: The View from Earth

In this chapter you have read and learned some about the moon and about scientific explanations. The investigation you just did demonstrated a model of the apparent changes in the shape of the moon.

Part A

Individually read through the following explanation and decide whether or not this is a scientific explanation. Record and justify your decision in your notebook.

We see phases of the moon because the earth and the moon do not sit still in space. They are both moving around the sun, which is a constant source of light. The sun shines on both the earth and the moon. When we see moonlight, we really are seeing the sunlight that is reflecting off the surface of the moon. When we see the lighted side of the moon, we are seeing its "daylight" side. When we see phases of the moon, we are seeing parts of the moon's daylight side.

The phases of the moon have the shape they do because the moon is a sphere (like a globe) and because of our location on the earth when we view the moon. Half of the moon is always lit, but we cannot always see all that is lit because the lit part of the moon is not always directly facing the earth. When the moon is almost between the earth and the sun, we see the light that is reflected off the moon where it is coming around the curved side of the moon. Look at Figures 4.2 and 4.3 for examples of how the moon appears at different times in its cycle.

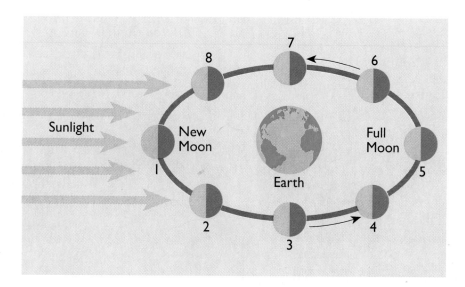

Figure 4.2

These are the eight major views we see of the moon. This illustration shows where the moon is, compared to the sun and the earth, for each phase.

Figure 4.3

This illustration shows how much of the moon is lit for each phase.

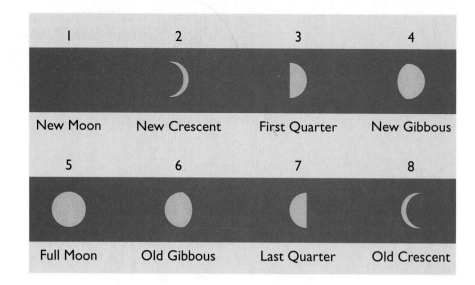

1	2	3	4
New Moon	New Crescent	First Quarter	New Gibbous

5	6	7	8
Full Moon	Old Gibbous	Last Quarter	Old Crescent

Actually the earth, the moon, and the sun aren't always lined up. If they were, then we would see an eclipse of the moon, as shown in Figure 4.4, every month. Partial eclipses happen, however, only about twice a year.

Part B

Draw a picture of the positions of the sun, the earth, and the moon when we see a new moon.

Once you understand the scientific explanation for the phases of the moon, your predictions about what the moon will look like will be more accurate. For example, if you know the exact relationship of the earth, the sun, and the moon as they move, you can predict exactly when eclipses will occur. If you didn't have a

Figure 4.4

This is the position of the sun, the earth, and the moon during an eclipse of the moon. The earth casts a long shadow into space. When the moon passes through the earth's shadow, an eclipse of the moon occurs. A total eclipse of the moon happens at least once a year, on the average.

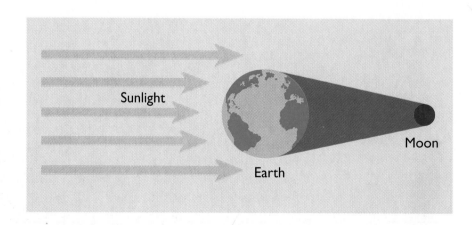

Sunlight

Earth

Moon

scientific explanation of the moon's phases, you would have to observe the moon for a long time to figure out the pattern of eclipses. Now you know how explanations that are scientific can help you make predictions about patterns. Furthermore you can improve the accuracy of your predictions each time you improve your scientific explanations.

Recognizing Patterns of Change

Look at the trees in these photographs. What patterns do you recognize? How could you explain these patterns to someone else? The world is full of patterns and many patterns help scientists explain changes that we all experience. Use the questions in this chapter to evaluate your understanding of patterns of change.

READING:
Can You Recognize Patterns of Change?

In this unit you made observations, recognized patterns, and made predictions in order to develop scientific explanations. For example, you and your classmates began watching the moon some time ago. Your observations and drawings helped you recognize the pattern of the moon's phases. Then you tried to explain the pattern you saw.

Rosalind's suspicion is correct. A big idea does connect the subjects in this unit. The moon, plant growth, and foot length all share the common thread that Rosalind thought of: They all display patterns.

Think back to your plant growth experiment. You watched these plants and recorded the pattern of how quickly they sprouted and how they grew. Because you had set up a controlled experiment, you could make predictions about future plant growth.

Back at the beginning of the unit, you measured foot length. You probably found a pattern in foot lengths, and if you looked back in your notebook, you would be able to predict the most common foot lengths for people your age. Because humans show a pattern of growth, you could make predictions about the foot lengths of people who are younger than you are or older than you are.

Stop and Think

1. So just what is a pattern?
2. What patterns can you recognize?
3. What patterns can you explain?
4. What is the pattern between the type of activity you are doing (reading, connections, or investigation) and the symbol next to the title?
5. What patterns do you see in the photographs at the beginning of Chapters 1, 2, 3, and 4?

 CONNECTIONS:
Putting the Pieces Together

Included in each of the following paragraphs is at least one question or challenge. As you review this unit, make sure you can answer each question or challenge.

Throughout this unit you have been identifying patterns of change. In the first chapter you solved puzzles, measured feet, observed pendulums, and made music. Think back to Chapter 1 and be sure that you can tell someone else what a pattern is and that you know the differences among cycles, trends, and correlations.

In Chapter 2 you started your plant experiment. You might remember details about Captain Sneezy's or about the bean plants, but can you identify the factors that might affect the growth of a plant? Can you design an experiment to test the effectiveness of a medicine? Can you design a controlled experiment?

In Chapter 3 you analyzed graphs in the investigation Finney's Funny Food, observed an unusual trend about magnets, and made predictions about the patterns of objects that sink or float. The main idea to remember from these investigations is that accurate predictions require information of sufficient quality and quantity. Can you give examples of times when you used information to make a prediction? Did you have enough information? Was your information of sufficient quality?

In Chapter 4 you looked at legends and wrote stories that explained patterns. Then you looked at the moon watch records that you had kept. This should have helped you notice a pattern about the shapes of the moon. You read about how some people, using legends or stories, have explained this pattern through time. Another type of explanation is a scientific explanation. Can you explain the difference between a story and a scientific explanation? How did you use the moon watch experiment to connect patterns, predictions, and scientific explanations?

INVESTIGATION:
What's Going On Here?

Several hundred years ago people died from diseases people rarely die from today. People had to face serious illnesses with little hope of a cure. In this investigation see whether you can use your knowledge of patterns to make predictions. Maybe you can discover what would have kept some people from getting sick.

Working Environment

You and your partner will join with another team of two to work in a cooperative team of four. You will all be Team Members, and you will need a Tracker and a Communicator. Push your desks together to form a table or sit close together at a table. Practice the social skill Speak softly so only your teammates can hear you.

Procedure

1. Get into your teams.
2. Listen to your teacher read the guided imagery.

   Make your notebook entry. Include today's date and investigation title.

3. Read aloud the descriptions of people in the village of Rainy Corner.

   Each Team Member should read one description. Then the Communicator and Tracker each might read one more. Listen politely as your teammates are reading.

Descriptions

Colin Smith—Blacksmith, Farmer

   Colin is 28 years old and has five children. He always has been strong and healthy. He enjoys working outdoors but also works making horseshoes and plow blades in his blacksmith shop.

Ann Howard—Seamstress

   Ann Howard has traveled to London and is a fine seamstress. She sews gowns, curtains, and everyday items for the women of the village. She is 30 years old, has three children, and is married to James Howard. Unlike most people in the village, she does not keep a garden because she does not enjoy being out in the cold, damp climate.

Ellen Throckmorton—Milkmaid

   Ellen is 20 years old and works for the Smith family. Her duties include cleaning the house, watching the children, and milking the cows. Ellen milks the cows twice a day. She is actually fond of the cows; she calls the cows by name when she takes them to pasture.

James Howard—Farmer

James Howard is 35 years old and farms the land next to Colin Smith. He has always enjoyed farming, even in the damp climate of England. He is married to Ann Howard, and they often visit the Smith family on Sundays. James has always had good health, except for a childhood injury to his leg. He now walks with a slight limp. He enjoys his work and takes care of the cows, chickens, and pigs, as well as his fields.

Sarah Wesley—Hired Girl

Sarah Wesley, age 19, works for the Howard family and is a friend of Ellen Throckmorton. She often visits Ellen when Ellen is taking care of the Smith children. Sarah prefers to be indoors most of the time, and she enjoys working for the Howard family. She is glad that she does not have to do the outdoor chores that Ellen has to do.

Richard Cooke—Baker

Richard Cooke, age 26, sees all the people of the village. He is famous for the good bread he bakes—even visitors from London stop to buy it. He enjoys being a baker because he dislikes working outdoors in the cold. He does not yet have a family, but has been known to visit Ellen Throckmorton while she works at the Smiths. He dislikes the barns, but he occasionally helps Ellen with the farm chores anyway.

4. Compare the descriptions.

   To compare descriptions, discuss the following:

   ■ *What do the people have in common?*

   ■ *How are their situations different?*

5. Draw a data table in your notebook that looks like Figure 5.1.

   Be sure your data table has six rows.

PREDICTIONS

Person's name	Will get sick	Won't get sick	Reason

Figure 5.1

Use a data table like this one to record your predictions about who will or will not get sick. Copy this data table into your notebook.

6. Record each person's name in the first column of your data table.

   We described six people, so you should have six names.

7. Discuss with your teammates who you think will or will not get sick. Then record your predictions for each person.

   Place a check mark in the appropriate column.

8. Write a reason for each of your predictions in the last column.

9. Check that all Team Members are ready to present your team's predictions to the class.

10. Participate in the class discussion.

Wrap Up

After participating in the class discussion, record the answers to the following questions in your notebook.

1. How accurate were your predictions?

2. What patterns did you use to make your predictions?

3. Describe what people could have done to keep from getting sick.

4. How did you express your thoughts and ideas within your team and during the class discussion?

5. In teams of two you have been practicing the social skill, Speak softly so only your teammates can hear you. What did you have to do differently when working in teams of four?

READING:
The Value of Patterns and Scientific Explanations

Throughout history people have faced situations as puzzling as the one you just encountered. Once you knew that the milkmaids never got sick, you had a clue—an important piece of the puzzle. But this clue was not the whole story. You needed more information of a different type to discover that the cows often had a milder form of smallpox called cowpox.

This discovery gave you a second piece of the puzzle. If you were able to link these two pieces of the puzzle, you might have concluded that milkmaids got this mild form of the disease by touching the cow's udder.

This new clue eventually might have lead you to the realization that catching cowpox was a *good* thing because it made people immune to smallpox. Once you understood this, then you would have to figure out how everybody could catch cowpox so that they wouldn't get smallpox. It took more than 160 years from the point when people first realized that milkmaids never got smallpox to the point when people figured out how to protect the whole world from smallpox.

The fact that milkmaids got cowpox but not smallpox was a pattern. Based on this pattern, you could make a prediction that milkmaids would not get smallpox. Determining the cause behind the pattern, however, is part of developing a scientific explanation. Once someone develops a scientific explanation, then that person or others could make an even more powerful prediction. In this case scientists were able to determine how the disease could be prevented in everyone. The scientific explanation was that cowpox prevented people from getting smallpox—if a person got cowpox, somehow he or she would not get smallpox.

Using scientific explanations, people have been able to eliminate or control other medical problems. A good example comes from England of the 1830s. Many people lived in wretched conditions. They lived in the streets and drank water contaminated with sewage. Water purification systems were unknown, and few sewers existed. During the summer of 1831, an epidemic of cholera (**KAH** ler uh) swept through England. Cholera caused people to have a very high fever and severe diarrhea; many died within a few days. At that time no one knew how to prevent it. No one knew what caused cholera or how it was transmitted. It struck rich neighborhoods as well as poor. Three people who observed and recognized patterns were responsible for saving thousands of lives.

William Farr began a study of all the afflicted neighborhoods in London. But he could see no pattern. Then after nearly 5 years, he saw a pattern that no one else had been able to see. Neighborhoods that were closest in elevation to the Thames River had the highest percentages of death due to cholera. Neighborhoods up on hills higher than the river had fewer deaths due to cholera.

Farr believed that cholera was caused by the stench from the river. Nearly 20 years later a doctor named John Snow looked at the same pattern in a different way. Dr. Snow thought that cholera was transmitted by polluted water. The pattern Dr. Snow observed was that the death rate was 10 times greater in areas where people obtained their water directly from the Thames than in areas that had a different water supply. He also observed that in a previously healthy district, 600 people suddenly died when a cesspit overflowed into the district well. This supported John Snow's explanation: Water polluted with sewage was causing cholera. John Snow could then predict that if a neighborhood had clean water, the people would probably not have cholera. A medical officer in London named John Simon worked hard to pass laws that required the city of London to clean up its water supply. Finally in 1858, city sewers were constructed and the cholera epidemic disappeared from England.

Now that you have identified several patterns on the surface of the earth, you can give an explanation for the patterns. You will be using a think-pair-share strategy to develop this explanation.

THINK Look over the list of questions you and your classmates just created. Also ask yourself, What do all the patterns have in common?

PAIR Pair up with your teammate and discuss what you each thought about the patterns. How did the patterns form? Between the two of you, try to explain the locations of the patterns on the earth.

SHARE Share your explanation with the rest of the class. Be sure to list the evidence you needed to develop your explanation.

The World Map

How long did it take to get an accurate map of the earth? It took more than 800 years! Al-Biruni, whom you read about in the play, was able to add map measurements on his maps only once a year, on the summer solstice (the longest day of the year). His mapmakers knew that the longer this day was, the farther north they were. They then could compare it with the length of day at their own home near the equator. Al-Biruni's assistants worked to improve maps that had been drawn several hundred years earlier. Four hundred years later, in the 1400s, the Europeans learned about the North American and South American continents, so the world map changed again—that is, mapmakers had more work. When the entire earth had to be mapped from the ground (instead of from space, as we do today), mapmaking took a very long time.

Connecting the Evidence

As you might have realized after reading the play in Chapter 7, sometimes explanations do not occur to scientists immediately. Sometimes scientists ask the same questions for hundreds or even thousands of years. But sometimes, when scientists cooperate and bring their ideas together, an explanation appears that answers many questions all at once. As you begin Chapter 8, see whether you can recognize questions that we are asking again. Later in the chapter, you will have the opportunity to see how these questions fit together and what scientists currently think the explanations are.

INVESTIGATION:
Desks on the Move

Something mysterious has happened in your classroom. Do things look exactly as they did yesterday? What are the signs that something has occurred, and what puzzle needs to be solved?

Here is part of the puzzle: Although it may seem impossible, your classroom has moving desks. Apparently they have been moving for years and have been holding secret meetings at night. Usually a teacher comes in early to straighten up the desks and clean up the evidence, but this morning your teacher left the evidence intact. What evidence can you find for how the desks were arranged last night?

Working Environment

In this investigation your entire class is to cooperate as you share information, so everyone is a Communicator. Also you are all to function as Team Members collecting evidence. Practice the unit skill Show respect for others and their ideas.

Materials

For the entire class:
- clues

Procedure

1. Find out how the desks were arranged when they had their meeting last night.

2. Tell the class recorder about clues you discover.

3. Decide how to interpret the evidence and discuss the evidence with your classmates.

   Examine the list of clues. You and your classmates might not agree on what the evidence means, so be sure to discuss different opinions as they arise.

4. After the class agrees on a method of arranging the desks, group the desks according to how you think they were arranged last night.

5. Sit at your desk or table in its new location.

Wrap Up

By yourself, write answers to the following questions and then discuss your answers with your classmates.

1. Write several sentences to summarize the evidence you found.

2. What conclusions did you draw from the evidence?

3. Draw a picture of the way you think the desks were arranged during their "meeting" last night.

4. How did you apply the social skill? Describe two specific instances.

5. Assign a class rating for using the social skill. Since the beginning of Unit 2, as a class our use of the unit skill is: much improved, improved, slightly improved, unimproved.

INVESTIGATION:
Where Have the Continents Been and Where Are They Going?

During the previous investigation, you probably realized that certain colors and shapes of paper gave you clues about which desks might have been situated near each other. But you might be puzzled about why you studied moving desks in science class. (You also probably had a healthy skepticism about moving desks.) Believe it or not, your moving desks have something to do with maps, and they also might help you develop an explanation for earthquakes and volcanoes.

Materials

For the entire class:

- 1 large world map

For each team of two students:

- 2 pairs of scissors
- 1 copy of the Land Mass Worksheet
- 1 sheet of light-colored construction paper
- 10 cm of transparent tape or a glue stick
- 1 copy of Additional Information (for Part B)

Procedure: Part A—Looking at the Land Masses

1. Read through all the steps of this procedure.

2. Pick up the materials.

3. Cut out the land masses from the team worksheet.

Divide the cutting task fairly.

4. Arrange the land masses on a piece of construction paper in the way they are arranged on the large world map.

DO NOT glue or tape them down yet!

5. Find as many patterns in the shapes of the continents as you can.

Notebook entry: Record these patterns in your notebook.

6. Read this information.

North and South America are moving away from Africa at a rate of 2.5 cm per year. In other words, 10 years ago they were 25 cm closer to each other than they are now (2.5 cm per year x 10 years = 25 cm), and 100 years ago they were 250 cm (8 ft) closer than they are now.

7. Think about these ideas.

a. The continents have not always been as they are now.

b. We can see patterns on the map of the world.

8. Arrange the continents in the way you think they were 200 million years ago. Draw this arrangement in your notebook.

9. Write down the evidence that led you to arrange the continents in the way you did.

   Notebook entry: Record this evidence.

Procedure: Part B—Additional Information

1. Pick up the Additional Information Sheet.

2. Study the sheet.

3. Change the arrangement of the continents from 200 million years ago as necessary to match the new evidence that you have from the Additional Information Sheet.

   If you have trouble deciding how to use the new information, think back to the investigation Desks on the Move.

4. Tape or glue your new arrangement onto the sheet of construction paper.

5. Again draw in your notebook how you now think the continents were arranged 200 million years ago.

Wrap Up

1. In Chapter 7 both Antonio Snider and Alfred Wegener told the characters to look at a map. What pattern do you think each of them saw on a world map?

2. How did you use the new evidence in Part B to change the arrangement of the continents?

3. Do you think the continents really could have moved? Explain your answer.

4. If you feel that your team deserves a reward for staying with your group, think of one with your teammates and tell your teacher what it is.

READING:
Combining Ideas

You began this chapter by using evidence to figure out where your moving desks were the previous night. You just did something similar in the investigation Where Have the Continents Been and Where Are They Going. You probably looked at the shapes of the continents and noticed that some of the edges matched. This evidence, along with your information that the continents presently are moving, might have helped you figure out what the continents looked like a long time ago.

   After you cut out the continents and arranged them, you observed patterns. But you still might have had questions. Why would the continents have moved? A particular group of scientists, called geologists, also have this same question. Groups of geologists have studied maps, volcanoes, or fossils, and all have had questions after they observed certain patterns. But they could not explain the patterns. Each group of scientists had evidence, but they did not know how to pull the evidence together for an accurate explanation.

Alfred Wegener, a scientist whom the characters met during the play in Chapter 7, observed patterns and gathered evidence. Wegener thought that because the shapes of the continents fit together like a puzzle, it was likely that the continents once had been together and later moved to their present locations. Wegener called this theory **continental drift**. Other people supported Wegener's basic idea.

Wegener and other scientists found further evidence that the continents once might have been together. For example, geologists found **fossils** (FAH suhls) in Africa and Australia that were similar to fossils in South America and Antarctica. Fossils are the hardened remains or traces of animals or plants that lived long ago. Other geologists found that some fossils in North America and Europe also matched. This evidence, that similar animal and plant fossils exist on different continents, suggested that the continents had at one time been together but then moved apart.

Often the types of animal and plant fossils that scientists found did not match the present-day climate of the continent. For example, scientists found tropical plant fossils on ice-covered Antarctica. This evidence seemed to indicate to some geologists that once the continents were much closer and more similar.

But many scientists wondered how something as heavy as a continent could actually move. In other words, they had found a pattern on maps, as well as a pattern left by fossils, but they could not explain why the pattern occurred.

Another group of geologists identified other patterns on the earth. They knew about the locations of earthquakes but did not know why the earthquakes occurred where they did. They had noticed that earthquakes occurred in tight clusters in certain locations. Earthquakes were most common near mountain ranges and in certain locations in the ocean. Within the ocean many of these earthquakes occurred in lines along the ridge tops of undersea mountains.

But geologists still had more patterns to explain. People who studied volcanoes had observed the locations of volcanic eruptions. They found that, like earthquakes, the locations of volcanoes form a linear pattern—they occur in lines. They are common near mountains and along the ridge tops of undersea mountains. The pattern was evident, but geologists could not explain why it occurred.

Geologists finally shared their ideas, just as you shared your ideas in Desks on the Move. They combined their evidence to develop an explanation not only for earthquakes and volcanoes but also for other geologic events as well.

Geologists concluded that the surface of the earth is moving. But they explained that it is not just the continents that are moving. Geologists continued to explain that the continents sit on top of **plates** that make up the entire surface of the earth. It is these plates that move (see Figure 8.1). We call this explanation the **theory of plate tectonics**. According to this explanation, as the plates rub against each other, they cause earthquakes. Volcanoes occur where the plates move apart or where one plate pushes and melts beneath another plate. This also explains why the rocks are youngest in the middle of the Atlantic Ocean. Along the Mid-Atlantic Ridge, two plates are pulling apart and new lava is rising up between them. As this lava cools, it forms the youngest rocks.

Figure 8.1

Scientists have identified the major plates on the earth's surface. The continents move because they are part of these plates.

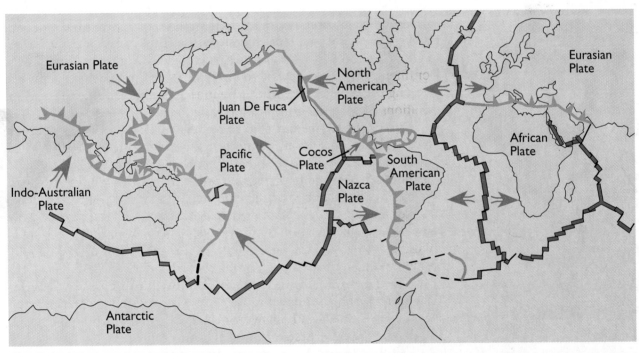

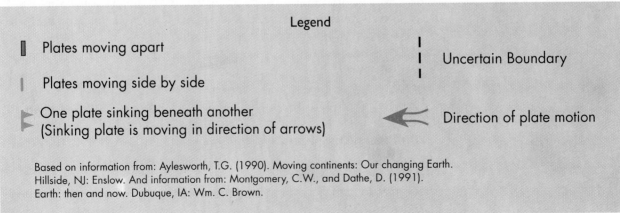

Legend

▌ Plates moving apart

▏ Plates moving side by side

◄ One plate sinking beneath another
(Sinking plate is moving in direction of arrows)

▮ Uncertain Boundary

◄ Direction of plate motion

Based on information from: Aylesworth, T.G. (1990). Moving continents: Our changing Earth. Hillside, NJ: Enslow. And information from: Montgomery, C.W., and Dathe, D. (1991). Earth: then and now. Dubuque, IA: Wm. C. Brown.

1. List some of the patterns that scientists noticed but did not know how to explain.
2. How does the theory of plate tectonics explain earthquakes?
3. According to the theory of plate tectonics, why are the youngest rocks in the Atlantic Ocean in the middle of the ocean?

What makes the idea of moving plates a strong scientific explanation? Should we accept it more readily than we might accept Aristotle's idea? If you remember from the play, Aristotle was a great thinker. He made careful observations, asked questions, and proposed that wind causes earthquakes. Today the idea that moving plates cause earthquakes is accepted more widely than Aristotle's explanation. Primarily this is because the evidence for plates is more convincing than the evidence for wind-caused earthquakes. The evidence includes the locations of earthquakes and volcanoes, the ages of rocks, and the fossil evidence from the continents. Plate tectonics also *connects* many different events—the moving of continents, the locations of earthquakes, and the locations of volcanic eruptions.

Furthermore, the theory of plate tectonics successfully predicts new observations. For example, the theory of plate tectonics explains how earthquakes and volcanoes are related. If small earthquakes begin to happen in an area that also has volcanoes, then we can predict that a volcanic eruption will occur. The stronger the earthquakes, the more likely that an eruption will occur soon. We also can predict that places far from plate boundaries are much less likely to have earthquakes. Predictions of this sort have provided additional support for the theory of plate tectonics.

The theory of plate tectonics is a strong scientific explanation. It is a strong explanation because it uses all the components of a scientific explanation. It answers questions, it is based on evidence, and scientists have tested it. Furthermore it connects observations and allows for predictions. Moving continents, volcanoes, and earthquakes all make more sense when we explain them in relation to the idea that giant plates of rock move across the surface of the earth.

Before you go on, think for a minute about the explanations that *you* developed for earthquakes and volcanoes in Chapter 7. Do any parts of your explanations agree with the theory of plate tectonics? Do you think that you would have come up with the theory of plate tectonics if you had had all the evidence? No matter what your answer, you should realize that just because plate tectonics is the currently accepted theory, it is not necessarily completely correct. Maybe some of your ideas or someone else's could help improve the theory.

4. How does the theory of plate tectonics explain why the continents are moving?

5. How does plate tectonic movement explain the formation of volcanoes?

6. Use the theory of plate tectonics to explain the correlations you saw between the locations of volcanoes and earthquakes.

CONNECTIONS:
Putting It Together

The previous reading provided you with a lot of information. One thing it said was that the continents moved because they are on top of moving plates. To see how the continents' pattern would have changed through time, assemble the flip book called Pangaea. Your teacher will hand out the scissors and the copies of Pangaea. See how Pangaea's pattern has changed during the last 200 million years.

INVESTIGATION:
But Why Should They Move?

I've never seen plates moving, so how am I supposed to accept this theory?

Explain ■ *Elaborate*

It might seem that the theory of plate tectonics presents a complete explanation of earthquakes and volcanoes. But some questions are still unanswered. If the earth really has plates, what are they made of? What do they sit on? And most important, why would they move? Our explanation will be stronger if we can answer these questions. In this investigation you will determine how solid plates might move.

Working Environment

Work cooperatively in your teams of two. One of you will need to be the Manager. As you work, practice the skill Stay with your group. Also watch out for each other's safety because your working environment includes a heat source.

Materials

For the entire class:
- 1 clock

For each team of two students:
- one 500-mL glass beaker
- 400 mL of water at room temperature
- 6 squares of card stock (2 cm square)
- 1 dropper bottle with food coloring
- 1 ring stand (or other apparatus to support the beaker)
- 1 heat source (Bunsen burner or votive candle)
- matches
- 2 pairs of goggles
- 1 pair of tongs

Procedure

1. Collect the materials.
2. Fill the beaker about 2/3 full of water and set up the materials as shown in Figure 8.2.

Figure 8.2

Use caution when you set up this equipment. Make sure things are steady before you add water to the beaker and before you light the heat source.

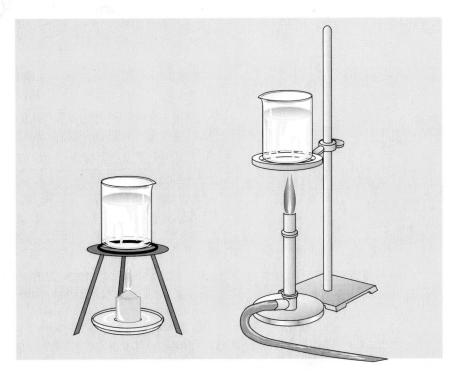

3. Put on your goggles and heat the water.

   If you use a Bunsen burner during this part, keep the flame very low. If you are not experienced with this equipment, see How to Use a Bunsen Burner (How To #7).

> ▲ CAUTION: **Hot liquids can cause burns.**

4. When the water begins to give off steam, add 2 drops of food coloring.
5. Observe what happens for at least 2 minutes.

   Notebook entry: Record your observations.

6. Just as the water begins to boil, use the tongs to add the paper squares to the water.
7. Observe what happens as the water boils more strongly.

   Notebook entry: Record your observations.

Background Information

The earth consists of more than the rocky surface that we walk on. In the past scientists have called the outside shell of mostly hard rock the **crust**. Currently, they refer to this as the **upper mantle**. Beneath that is a softer layer they call the **lower mantle**. To imagine what the mantle would feel like, think about a bowl of gelatin; it's not hard, but it's not exactly liquid either. The very innermost part is the **core**, and it is thought to be extremely hot. The core is made up of a liquid outer core and a solid inner core. The theory of plate tectonics focuses on the crust and upper mantle. The plates are made of the earth's crust and uppermost mantle. Plates are up to 80 kilometers deep on the continents and as thin as 5 kilometers beneath the oceans.

What is the earth like beneath the plates? Because of volcanoes, we know that somewhere inside the earth there is molten, hot liquid rock. In fact, scientists think that much of the interior of the earth is semisolid—more like a bowl of gelatin than the solid rocks we walk on. When molten rock reaches the surface, as it does during volcanic eruptions, it cools and hardens, just as water hardens into ice when you freeze it.

You might wonder whether or not anyone has drilled a hole to see whether these ideas are right. Unfortunately no one has been able to—yet. From the crust to the center of the earth is more than 6,400 kilometers, and we only can dig down 10 kilometers or so with our modern drilling equipment. Scientists have gathered their information about the earth's interior by observing how earthquake tremors travel through the earth and arrive at different places at different times.

Figure 8.3

Traditionally the earth has been divided into three parts: a crust of hard rock, a mantle that is softer (more like soft gelatin), and an inner core. The theory of plate tectonics focuses on the crust and upper mantle. Earth scientists have discovered that plates include slightly more than the crust because a thin slice of the uppermost mantle moves with each plate. Plates can be as thick as 80 kilometers. These thick plates include 65 km of crust and 15 km of the upper mantle.

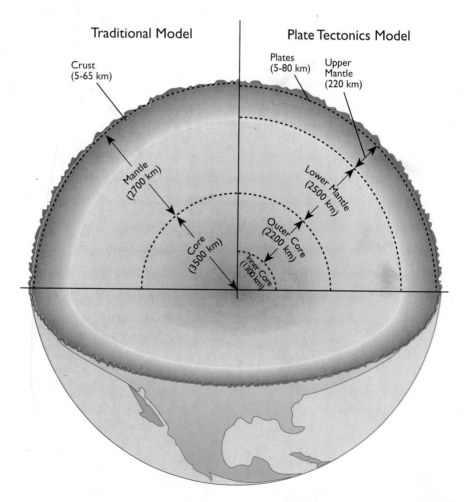

Traditional Model | Plate Tectonics Model

Crust (5-65 km)
Plates (5-80 km)
Upper Mantle (220 km)
Mantle (2700 km)
Lower Mantle (2500 km)
Core (3500 km)
Outer Core (2200 km)
Inner Core (1300 km)

Wrap Up

Participate in a class discussion of the following questions. Then write or illustrate your answers.

1. What did the paper squares represent?

2. What did the hot water and the heat source represent?

3. When you added food coloring to the warm water, what pattern did you observe?

4. When you heat water, a pattern occurs. Explain why molten rock beneath the plates could cause the plates to move. Base your answer on the pattern you saw.

5. Look at Figure 8.3 in the Background Information. Draw a side view of an oceanic plate and a side view of a plate with a continent on it. Label the thicknesses of each of these plates.

6. Why do you think plates with continents on them would move more slowly?

7. Explain how molten rock can become solid rock.

8. Why could rock on the ocean floor be recycled many times?

9. How might liquid beneath a solid rock plate help the plate move?

10. List specific strategies your team uses to stay together as a group.

INVESTIGATION:
Near the Edges

Where does most of the action take place on the planet? If you answered that most of the action takes place near the edge of a plate, you are right. Even though it might seem that plates easily slide past each other, quite a lot of action takes place at the edges of plates. Sometimes plates slide past each other, sometimes they move away from each other, and sometimes they move toward each other. These three types of motion—beside, away from, or toward—result in different types of boundaries. In this investigation you will be working with teammates to describe what happens at different types of plate boundaries.

Materials

For each team of two students:
- 1 stick of modeling clay
- 1 paper towel, cut in half

Working Environment

Work cooperatively in teams of two until step 5 and then meet with two other teams of two. You will be in a jigsaw group—one that divides up a task among its members. As you work in teams of two, use the roles of Communicator and Manager. Throughout this activity try practicing the new skill Treat others politely.

Procedure: Part A—The Social Skill

1. With your teammate think of two reasons why treating others politely is an important skill in cooperative learning.

2. Record your ideas.

3. In your notebook construct a T-chart for this skill.

Procedure: Part B—Plate Boundaries

1. Get the materials.

2. Read the background information section assigned to your team.

   These readings are located after the procedure.

3. Divide the modeling clay into two pieces.

   The Communicator and the Manager each should have one piece of clay.

4. Decide how your team will demonstrate your type of plate boundary to two other teams.

5. Get into your jigsaw group.

   Share your T-charts for treating others politely.

6. Demonstrate your team's boundary to the other members of your jigsaw group.

7. Be ready to demonstrate any of the three types of boundaries during a class discussion.

Background Information

Ridges

The ocean floor makes up about 71 percent of the earth's surface. For centuries people thought that the ocean floor was almost entirely flat. They had no evidence to the contrary because no one could explore the deep ocean waters. Extreme darkness, icy cold, and great pressure near the ocean floor prevented exploration. During World War II, however, scientists developed a technology to map the seafloor—special echo sounders and deep-sea cameras. The echo sounders allowed sailors to measure the depths of ocean waters and to determine where undersea mountains and canyons were. This work continued into the 1970s. Naval scientists then had enough information to make an accurate map of the ocean floor.

The map shows that parts of the ocean floor are flat. It shows that the ocean floor is make up of wide valleys, deep trenches, and towering mountain chains, called **ridges**, which are as high

Figure 8.4

At a ridge two plates are splitting apart, and molten volcanic rock (lava) is rising up in the middle.

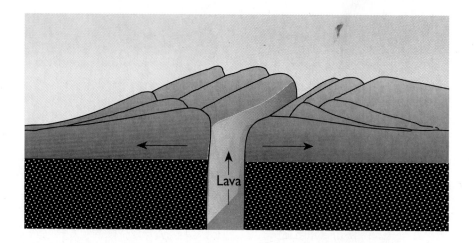

as 2,400 meters (7,880 feet). Figure 8.4 is a diagram of a ridge on the ocean floor.

Underwater ridges are the longest mountain chains on the earth. They wrap around the earth like the seams on a baseball. In some places the ridges are 4,800 kilometers (3,000 miles) wide. This is wider than the United States. Nearly all ridges are hidden under many kilometers of ocean water. Iceland and the Azores are the tops of ridges that rose out of the sea millions of years ago.

Most ridges are split down the middle by a valley. In the valley the edges of the plates are spreading apart. This spreading opens cracks in the valley floor. Hot **magma** (molten rock) oozes out of the cracks. When it reaches the earth's surface, it is called lava. The lava hardens into new mountains of rock. Like a conveyor belt, the mountains are moved away from where they formed. But once the mountains are pushed away from where they were formed, they do not get larger. The newest mountains are nearest to the valley; the oldest are the farthest away.

Trenches

Trenches are the deepest parts of the ocean. At a trench the edge of one plate bends and dives under another plate. The diving edge slowly moves down inside the earth. There, temperatures are hot enough to melt the rock. Sometimes the magma (molten rock beneath the earth's surface) may rise and break through the ocean floor to form volcanic islands. The islands of Japan were formed in this way. Figure 8.5 illustrates two different types of trenches.

Because one plate melts beneath another plate, trenches erase certain evidence. Scientists study rocks from the earth's crust to interpret earth's history, and some of these rocks have melted and disappeared. Rocks from the ocean floor that formed more than 200 million years ago have disappeared down into the trenches and have melted inside the earth. To obtain rocks older than 200 million years, scientists must study continental rocks. Continents do not go

Figure 8.5

At a trench one plate is sinking beneath the other. The sinking plate melts as it reaches the warmer temperatures inside the earth, and magma rises up to form volcanoes. Trenches can occur where oceanic plates meet (a) or where an oceanic plate sinks beneath a continental plate (b).

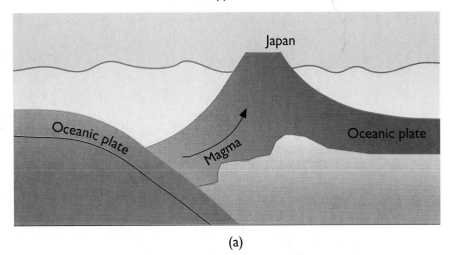

(a)

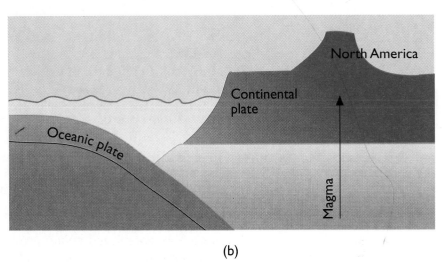

(b)

down into trenches. The rock is too light, and it essentially floats on top of the heavier rock below it.

Transform Faults

Transform faults occur when two plates slide past each other. When this movement occurs, it can produce earthquakes. The pressure between the plates builds until, suddenly, one plate moves past the other. This rapid release of movement and energy is an earthquake. Figure 8.6 shows a transform fault. One example of a transform fault is the **San Andreas fault** in southern California. When the Pacific and the North American plates move along this fault, earthquakes occur.

Some scientists think that transform faults may be the longest lasting type of plate boundary. After listening to the presentations of your jigsaw group, tell your group about transform faults. Explain why (for example) a transform fault might last longer than a trench.

Figure 8.6

At a transform fault, two plates are sliding past each other. This type of motion is what causes earthquakes in California.

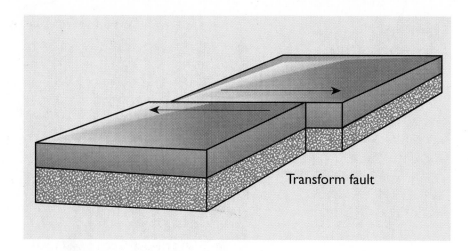

Transform fault

Wrap Up

Complete the following tasks in your jigsaw group.

1. Go back to Figure 8.1, the map of the plates on the earth. Draw a copy of Figure 8.1 in your notebook. On your drawing use the letters *R* (ridge), *T* (trench), and *F* (transform fault) to locate the ridges, trenches, and transform faults.

2. Decide what new strategies you could use to better practice the social skill of treating others politely. Record the strategies in your notebook.

INVESTIGATION:
Plate Tectonics Research

In this activity you will study one topic that you choose from a list. The list includes many events that are related directly to plate tectonics. You will become an expert on that one topic so that you can make a presentation to the rest of your class. As you listen to the other presentations, you will see how the theory of plate tectonics has explained many features on the earth.

Materials

For each student:
1 Presentation Review Sheet

Procedure

1. Choose a topic that you would like to research and record it in your notebook.

   You may choose a topic from the list at the end of this investigation, or you may choose a different one that your teacher approves.

Working Environment

You will work individually in this investigation. As you get help from others and as you listen to presentations, practice your unit skill and the skill Treat others politely.

Elaborate ■ *Evaluate*

2. Read any information in the Background Information that relates to the topic you chose.

3. Think about what else you would like to know about your topic and how you might find that information.

   You will need to visit the library in order to complete this investigation.

4. As you investigate your topic, do the following things:
   - Identify the question you're trying to answer.
   - Describe the answer you found.
   - Describe how the ideas connect with the theory of plate tectonics.
   - Use at least three references.

5. Research your topic.

   If you do not know how to do library research, see How To #6, How to Conduct a Research Project.

6. Design and organize a presentation for your classmates. Your presentation must include one of the following:
   - a 3-minute talk to the class,
   - a poster or demonstration large enough for the class to see, or
   - a two-page written description.

7. Show your teacher what you plan to do.

8. Make your presentation.

9. As you listen to other presentations, listen for the specific things listed in step 4.

   Follow the example on the Presentation Review Sheet for each person you listen to, and write a review of each presentation in your notebook.

Topics for Presentations

1. How does the landscape change after a severe earthquake?
2. How do earthquakes affect buildings?
3. How could people's lives change after an earthquake?
4. How are rocks such as basalt and granite formed?
5. What are the differences between pumice and marble?
6. Why is basalt the main rock in Hawaii but granite is the main rock in the rest of the mountains in the United States?
7. What is a tsunami (tsoo NAH mee), and how does it form?
8. Why are some rocks on land billions of years old but in the ocean the oldest rocks are only 200 million years old?
9. If most mountains are found on the edges of continents, how can you explain the location of the Ural Mountains?

10. Major earthquakes have occurred within the last 10 years in Mexico City, San Francisco, and the Philippines. Report on one of these earthquakes or another recent one.

11. Report on a relatively recent occurrence of volcanic activity, such as that in Japan, Hawaii, or the Philippines.

12. Choose one of the following mountain ranges: the Andes, Himalayas, Cascades, or Appalachians. Explain how that mountain range may have been formed according to plate tectonic theory.

13. Use the theory of plate tectonics to explain why Mt. St. Helens erupted.

14. According to recent research, the continent of Africa is splitting apart. Describe and show where this is occurring and what the evidence is.

15. Describe what a hot spot is and how scientists can use a hot spot on the ocean floor to indicate the direction of a plate's movement.

Background Information

This reference section contains information about some topics that are related to plate tectonics but have not yet been discussed in the readings.

Mountains

Mountains are formed in several different ways: by volcanic eruption, by folding when plates collide, and by movement along some types of faults. **Folding** means that the rock layers bend but do not break (see Figure 8.7).

Faults form when the rock is pushed to the point of breaking. In this case a broken block of rock may be pushed up, down, or

Figure 8.7

Folded mountains form when rocks have been squeezed from opposite sides and thus form folds.

Photograph provided by Dr. Carl Vondra, Iowa State University, Geology Field Station.

Elaborate ■ *Evaluate*

Figure 8.8

Mountains that have formed by faulting often have different types of rock next to each other. When movement occurs along the fault, the crust breaks. As a result different types of rock appear next to each other.

sideways. Faults generally are evident in mountains when two different types of rock occur next to each other. Movement along the fault has pushed one type of rock next to another type of rock (see Figure 8.8). Many times, mountain ranges form as the result of a combination of folding and faulting.

Volcanic mountains form when volcanoes erupt (see Figure 8.9). The islands of Hawaii, Iceland, and Japan are examples of volcanic mountains. These mountains formed over hot spots, ridges, or near trenches.

Rocks

Rocks form in several ways. One type of rock forms when magma rises near the earth's surface and cools to form igneous (IG nee us) rocks. If the magna cools very slowly, large crystals have time to grow, and a rock such as granite forms. If the lava is thrown from a

Figure 8.9

Volcanic mountains form when molten rock erupts from inside the earth, and then the molten rock cools and hardens.

volcano and cools rapidly, a rock such as andesite may form. A piece of andesite might weigh the same as a piece of granite, but the crystals in andesite would be tiny because the rock cooled so rapidly.

Other rocks form because sediments (such as soil and pieces of animal shells) become cemented or glued together. Just as a mixture of sand and clay can form a cement sidewalk, small particles of clay can dissolve in water and hold natural sediments together. We call these cemented rocks sedimentary rocks. Examples include limestone, shale, and sandstone.

If a sedimentary rock is in an area where mountains are forming, it may become deeply buried and subject to great heat and pressure. The heat and pressure at great depths can change the sedimentary rock. Limestone can change to marble, shale can change to slate, and sandstone can change to quartzite. We call rocks that change this way metamorphic rocks.

Volcanoes

When a volcano erupts, magma comes to the surface. Magma is molten rock that is beneath the surface of the earth. When magma reaches the surface, it is called lava. Volcanoes are located over openings in the earth's crust and the material from the mantle comes out through the openings. This material is a combination of liquid rock and gases. If the gases are trapped or if a volcano is made of a particular rock type, then an explosion might occur.

As the magma comes out, it separates into gas and lava. The lava cools rapidly and forms rocks such as basalt or pumice. Ash and cinders also come out of the volcano and build up around the sides of the opening.

Hot Spots

Some places on the earth have had a lot of volcanic activity and yet are not near a plate boundary. These are called **hot spots**. Two

Figure 8.10

As a plate moves over a hot spot in the mantle, a small section of the plate melts, and magma rises up through the opening and forms volcanoes. But because the plate slowly keeps moving, one volcano after another is formed, creating a chain of volcanoes such as the volcanic Hawaiian Islands.

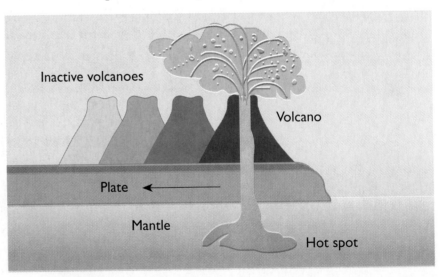

Elaborate ■ *Evaluate*

examples of places affected by hot spots are the Hawaiian Islands and Yellowstone National Park. Yellowstone National Park has a variety of volcanic rocks, hot springs, and geysers. A geyser (GUY zer) is like a small volcano. It erupts, but instead of hot lava, hot water and steam escape. The geysers are evidence that even if the rocks at the surface have cooled, there is still enough heat below to warm the water traveling through the ground.

CONNECTIONS:
Looking Back

Look back at the terms and phrases you listed in your notebook during the Chapter 7 investigation Patterns on the Earth. Which words and phrases are more familiar now? Which ones are still unfamiliar?

Also review the questions your class posted during the Chapter 7 connections activity Can You Explain the Observations? Which questions can you answer now? What new questions do you have?

In this chapter we have introduced you to the theory of plate tectonics. Earth scientists currently use this theory to explain many features on the earth. You are responsible for knowing the evidence that we have presented. But do you have to believe that moving plates cause earthquakes? No. Just because something is the currently accepted scientific theory does not mean it is correct. The evidence has been laid out for you, but you can still decide for yourself whether or not you agree with the conclusions presented in the chapter. Many scientists have disagreed with the current theories of their time. This is one way scientific advances take place.

CONNECTIONS:
Listen to How They Say It

At the beginning of Chapter 7, you looked at photographs and saw a film or video about earthquakes. In your class discussion, you probably discussed many ideas about what happens when volcanoes erupt and earthquakes occur.

Throughout this unit you have been reading about and coming up with your own scientific explanations. In this activity it's your turn to evaluate *someone else's* scientific explanation.

Before you watch the video or film that your teacher will show, review the readings in Unit 2. Think about the parts of a good scientific explanation. If you are hearing a complete scientific explanation, what things will the speaker tell you? What should you listen for? Write yourself a short list of the things you will listen for.

Listen and watch the video or film. How complete was the scientific explanation?

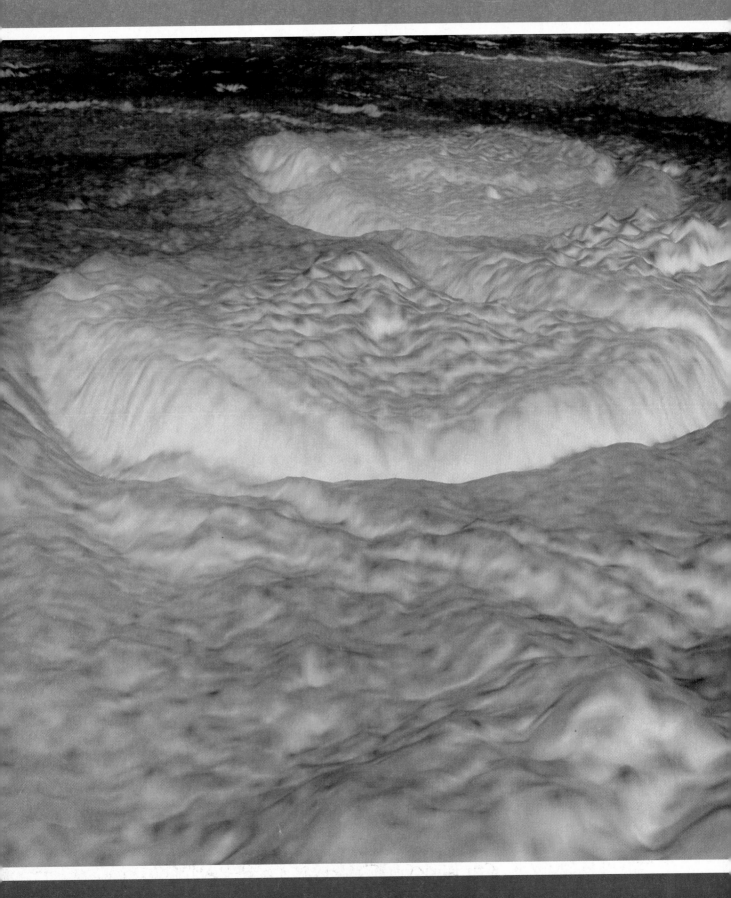

Using Scientific Explanations

This photograph is a color-enhanced computer image of the surface of Venus. These domelike hills may have formed due to the horizontal flow of lava. The image is assembled by the computer from data the *Magellan* spacecraft relayed back to the earth. How would you describe the surface of Venus? Do you see any evidence of plate tectonics on Venus? What would you look for to find such evidence?

READING:
Scientific Explanations for Experts

What have you learned about scientific explanations since Chapter 6? If someone were to ask you whether or not an explanation was really scientific, what would you check for?

Since the beginning of Chapter 6, you have gathered evidence to answer questions. At first the questions were very specific. You tried to discover the placement of the walls in a maze puzzle and to gather evidence for what was on the bottom of a cube. Next you used evidence to test your ideas about how an If–Then Box might work. As you tested ideas about the If–Then Box, you recorded your information and wrote if–then statements. Try to think of some if–then statements now. What examples can you give?

In Chapter 7 you looked at the patterns in the occurrence of volcanoes and earthquakes that people have observed. As you read a play, you saw that different people have observed patterns and have proposed explanations for volcanoes and earthquakes for thousands of years. Only recently, however, have people tested their explanations by gathering evidence *after* they proposed their explanation.

People can gather more evidence about earthquakes and volcanoes today because they have better tools than they did in the past. As you saw in Chapter 7, maps are one tool that took a long time for people to develop. For hundreds of years, people could tell how far they were from the equator only by measuring how long daylight lasted on the longest day of the year. After many years of measurements, people finally could draw maps that showed the shapes of continents. Remember, in those times no one could observe the earth from space. When people were able to draw more accurate maps of the world, they could observe more patterns about the earth. When you looked at earthquake and volcano data plotted on maps of the world, you saw patterns to their occurrence.

The fact that volcanoes and earthquakes occur most often in a linear pattern on certain parts of the earth has intrigued many people. After many years of individuals asking questions and proposing answers about the reasons for this pattern, scientists combined their ideas into a theory. They called it the theory of plate tectonics.

Scientists proposed the theory of plate tectonics in the late 1960s. Since then many scientific explanations have changed. Currently plate tectonics is the most accepted theory that explains why earthquakes and volcanoes occur where they do: Giant plates of rock float on the earth's surface, and beneath them lies extremely hot molten rock. Where these plates meet, earthquakes and volcanoes occur.

You might now know more of the specific facts about plate tectonics, but what is the purpose behind all of these activities? If you had chosen to memorize the locations of all the plate boundaries, you would have some good information, but you wouldn't have the whole story. If you really have learned the material in this unit, you will know how to recognize and propose a scientific explanation. If you can recognize and propose scientific explanations, you have learned how to analyze information, and that is a very important skill.

INVESTIGATION:
Plate Tectonics in Outer Space

The managers of a new company called Tectonics Research have asked you to work for them. They have planned a trip to the earth's moon and several of the earth's neighboring planets and their moons. The head of this company is convinced that if the geologists knew more about plate tectonics on other planets, they could do a better job of predicting earthquakes on earth. They want you to join them on their journey. They are not certain whether plate tectonics occurs on other planets and moons. They want you to help them decide which planet or moon would be the best place or places to study more about plate tectonics.

Working Environment

In this investigation you will work alone.

Materials

For each student:
- 1 copy of the Evidence Chart

Procedure

1. Read the Background Information that follows these procedures.

2. Participate in your class review of if–then statements.

   You will use if–then statements to test whether or not plate movements are occurring at the places you visit.

   "If plate tectonics is occurring, then I will expect to see mountain ranges in long linear patterns."

3. Think about the patterns you saw on maps of the world during Chapter 7. Then complete the following if–then statements for

   - earthquakes,
   - volcanoes, and
   - solid rock.

   For a and b in particular, think about the patterns you saw for earthquakes and volcanoes.

   a. If plate tectonics is occurring,

      then *(fill in with a statement about earthquakes)*

   b. If plate tectonics is occurring,

      then *(fill in with a statement about volcanoes)*

   c. If plate tectonics is occurring,

      then *(fill in with a statement about solid rock)*

4. During your journey, test your if–then statements. Write down all the information you have on the evidence chart your teacher will hand out.

5. Read through the following example.
   The spaceship has landed on the earth's moon. You have your protective spacesuit on, and you step out. You find a cold environment. You are walking in a flat and dusty crater with a few mountains around the edge of the crater. The rocks are volcanic. Your captain tells you that no one knows of any earthquakes here during the past 25 years of exploration.

6. This is your evidence. What will you conclude? We have filled in one possible example for you (Figure 9.1).

7. Read the description of Stop 1.

8. Fill in your data table.

   There will be specific evidence for you to test, but any conclusion you can justify will be considered correct.

EVIDENCE CHART

Stop	Results of first if–then (mountains in linear patterns)	Results of second if–then (earthquakes)	Results of third if–then (volcanoes)	Results of fourth if–then (solid rock)
Example: Earth's moon	Saw mountains.	No known earthquakes.	Saw volcanic mountains, but not in a line.	Saw volcanic rocks.
Stop 1 Io				
Stop 2 Mars				
Stop 3 Venus				

Conclusions	
Example: Earth's moon	We saw volcanic rocks, but most of the evidence does not indicate that plate tectonics is occurring today.
Stop 1 Io	
Stop 2 Mars	
Stop 3 Venus	

Figure 9.1

This example shows how you might fill in the Evidence Chart after a visit to the earth's moon.

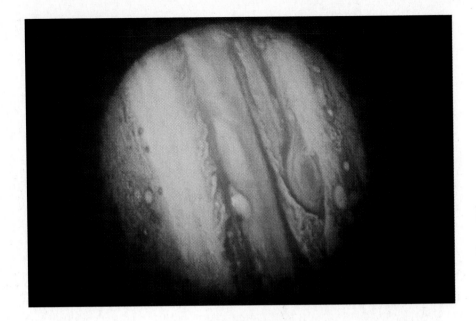

Figure 9.2

This NASA photograph shows the planet Jupiter. Jupiter is the largest planet in our solar system and has 16 moons. Io is one of them.

Stop 1

You land on a moon of Jupiter. This moon is called Io. When you climb out of the ship, you step onto a surface of bright yellow sulfur that appears to have flowed from a volcano. You can tell that this moon is extremely warm. Soon after you get out and begin to look around, you run back to the ship. In the distance you have seen a volcano beginning to erupt. As you run, you feel the ground shaking beneath your feet. Once you are inside, the ship takes off immediately. You ask the captain to orbit this moon. From above, you see many volcanic mountains all over the moon, and many of them have very obvious pools of lava spilling out. You cannot see whether there is any pattern to where they occur. You see mountains but no linear patterns.

9. Read the descriptions of Stops 2 and 3.

10. Fill in your data table for Stop 2 and then Stop 3.

Stop 2

This time you decide to have the captain orbit the planet before you land. You see Mars below. You see tall cones of volcanic mountains, but you do not see any in lines. None of the volcanoes appear to be erupting. Even looking straight down into what appear to be volcanic cones, you don't see any lava. When you ask about quakes, the captain tells you that no quakes or volcanic eruptions have been recorded in recent history.

When the spaceship lands, you look out onto a red and dusty planet. It's extremely cold. You see that the mountains you observed from space are very tall, and some probably are bigger than any on Earth. The rock is volcanic, and you decide to look once more for lava flows from the volcanoes. Even after a long walk and several different stops on the planet, you don't find any molten lava.

Figure 9.3

Like Earth, Venus is a planet that appears blue from space. Venus, however, has temperatures that are much hotter than those on Earth, and Venus has no moon.

Stop 3

As the spaceship approaches the planet Venus, it goes through thick layers of clouds. Even though you have planned to make observations from just above the planet, you can't see very clearly. Then the spaceship flies below the clouds, and you can see many volcanoes in long rows. On your first landing attempt, a quake occurs, and you have to take off immediately and try again. The captain is flying a straight course, and when you attempt to land again, the same thing happens—another quake. The captain then changes course, and after a careful approach, the spaceship finally lands. You can see mountainous rocks in the distance. Then the captain shouts that you must take off again because a wall of lava is approaching the spaceship. It is clear that the planet is very hot.

Background Information

Scientific evidence suggests that the planets in our solar system formed about 4.6 billion years ago. Initially the surfaces of planets were hot because they were constantly being struck by particles from space. Slowly more of the particles from space grouped together on the planets and enlarged them. The planets were no longer being constantly bombarded, and their surfaces cooled. Some of the planets cooled faster than others. In general, the farther from the sun, the cooler a planet is. Small planets tended to cool faster. The moons that orbit planets (such as earth's moon) usually are much smaller than the planets and tended to cool the quickest. There is an exception, however, which you will read about.

Wrap Up

1. Of the three locations you visited, which one or ones would be most suitable for studying plate tectonics?

Do You Know Your Neighbors?

Earth is one of nine known planets that move around the sun. Do you know your neighbors? If you don't, take a look at the diagram below. This will show you where the other eight planets are located.

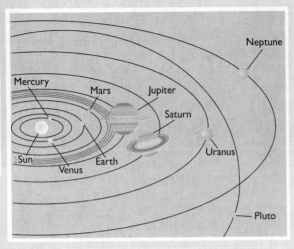

Each planet has an orbit around the sun. Usually in our solar system, Pluto is the farthest planet from the sun, but until 1997 its path will be closer to the sun than Neptune's.

2. For each place that you visited, explain why you think plate movement is or is not occurring at each of those places now.

3. Do you think that tectonic movement ever has been or will be active in any of the three places you visited? List the place(s) and explain why you do or do not think they have been or will be tectonically active.

INVESTIGATION:
The Mystery Planet

Working Environment

In this investigation you will work alone.

The theory of plate tectonics provides a strong scientific explanation for volcanoes, earthquakes, and mountains, in part because it allows us to make predictions. If we know the pattern and can explain its cause scientifically, we can make accurate predictions. In this activity see whether you can use the evidence we provide to recognize patterns and to make predictions.

Materials

For each student:
- 1 copy of the Mystery Planet Map

Procedure

1. Read the information that follows.

   The people at Time Travel, the company that gave Marie and her friends the free trip, are planning to develop several resorts. They have found a planet very similar to Earth but very young. It has cooled just enough to have formed a crust.

   Even though it will be about 200 million years before the planet is ready for occupation, this will be no problem for the people at Time Travel. If they want to, they can just go forward in time!

   But time travel is expensive. Before they send their builders on a trip to the future, they want to know where they ought to build. So they have asked you to help them.

2. Predict where they should build each of the following resorts.

   a. Snowy Mountain Ski Resort

   b. Magic Volcano International Park

   c. Stable Plains Horse-and-Family Dude Ranch

   d. Oceanic Ridge Deep Sea Diving Resort

3. Look at the Mystery Planet Map. It shows how the plates are moving.

4. On your map, label the best locations for these resorts.

Wrap Up

In your notebook explain why you chose the locations you did. Then draw a sketch in your notebook of a brochure that you would use to advertise one or more of the locations on the mystery planet.

I'm gonna be ready this time. No one will leave me behind for this time travel trip. Plan ahead I always say...

UNIT 3

Responding to Patterns of Change

By now you may feel that you have a fairly good understanding of what causes earthquakes and volcanoes. You looked at scientific explanations and the causes of certain events. But there are other events and situations that people need to respond to. As you can see in the photographs on these pages, sometimes people have to determine how they can survive dangerous storms. At other times people simply want to make the best use of their environment that they can. What can people *do* when they know that certain events may occur? In other words, how can knowing about patterns help us solve problems?

COOPERATIVE LEARNING OVERVIEW

The better you are at working in a team, the better able you are to use challenging skills. Think about it. You started by trying to express your ideas and listen politely. In the next unit, you practiced respect for the ideas that others were expressing. Now that you can express your ideas aloud and can respect the ideas of others, you can learn to encourage others to share their ideas. This will allow you to hear many different opinions. You also can encourage people who don't want to help to be a part of the team. You really are becoming a team player!

Now that you have come this far in cooperative learning, you probably have definite opinions about it. Some of you even might have found that working cooperatively is not limited to your science classroom. You might not work in teams in your other classes, but think of other group situations in which you cooperate. The characters have introduced you to your Unit 3 skill of encouraging others to participate. Before you begin Chapter 10, be sure that you have created a T-chart for this skill. Also share your feelings about cooperative learning with your new teammates. Be sure to express both the positive and negative aspects of cooperative learning that you have discovered.

You again will have social skills for each activity. Remember to use your unit skill in each cooperative activity even if you are working on a different activity skill. Also try to remember and keep using the social skills you worked on in past units. Finally review the role descriptions with your new teammates so that together you can answer any questions you might have about the duties of each role.

What Causes Weather Patterns?

Weather can be as calming as the soft sound of rain on the roof or as dangerous as the floodwaters from a hurricane. As you might know already, weather events can be as dramatic as earthquakes and volcanoes and sometimes just as hard to predict. People make attempts to predict the weather, however, because it can cause such hazardous situations. Weather forecasters can warn people about these hazards so that they can take action. For example, before a hurricane many people will listen to a forecast, along with the warnings and recommendations, and then may evacuate the area.

People can make predictions about weather because they've learned some of the factors that affect weather patterns. Can you explain how rain forms or how wind blows? By the end of this chapter, you should be able to do just that and describe the patterns related to the occurrence of weather.

INVESTIGATION:
Water on the Move

If you've ever been soaked in a rainstorm or even surprised by a thundershower in the desert, you know that water is a big part of the weather. Rain, snow, hail, and fog are all examples of water on the move. Even when you can't see it or feel it, water is moving around you. What evidence can you find in this investigation that water is on the move?

Materials

For each team of three students:
- 1 empty can, soup size
- water, room temperature ($\frac{1}{2}$–$\frac{2}{3}$ of a can)
- 3 ice cubes ■ 1 stirring stick ■ 1 paper towel

Procedure: Part A—The Social Skill

1. Write your teammates' first and last names in your notebook.
2. Create a positive two-line rhyme to go with each of your teammates' first names.

   Notebook entry: Record your name rhymes.

Procedure: Part B—A Chilling Experience

1. Obtain the materials for Part B.
2. Fill the can about half-full of room-temperature water.
3. Observe how the can looks and feels on the outside.

   Notebook entry: Record your observations.
4. Add three ice cubes to the can.

   The Communicator should do this.
5. Stir the water and ice slowly.

   The Manager should do this. Do not hold the outside of the can while stirring.
6. Keep stirring for 3 minutes.

   The Tracker should keep track of time.
7. Look at the can and touch the outside.

   Notebook entry: Record these observations.
8. Empty the can and dry it off. Then return your materials.

   Check that the outside of the can is dry.
9. If you haven't looked closely at it yet, observe the glass containers your teacher has set up.

   Notebook entry: Record your observations. Make sure that your observations account for the differences between the setups.

Engage ■ *Explore*

Wrap Up

Discuss the following questions with your teammates. Record your answers in your notebook. Each of you should be prepared to explain your answers if the teacher calls on you during a class discussion.

1. Explain what you think happened to the outside of your can.

2. How might you test the explanation you provided in question 1?

3. What similarities did you see between what happened with your can of icy water and what happened in the demonstration?

4. Write a two-line rhyme that describes how well you used each other's names as you worked. Combine this two-line rhyme with the name rhymes you created to make one short poem. As a team recite this poem to the rest of the class.

INVESTIGATION:
Wind in a Box

After one particularly long day at school, Marie arrived at home feeling exhausted, hot, and thirsty. She went to the refrigerator to

find something cool to drink. When she opened the refrigerator door, she noticed a particular phenomenon. What do you think happened? In this investigation you will have the opportunity to investigate this and other similar patterns.

Materials

For the entire class:
- 1 box of safety matches
- 1 large bucket of water
- 1 fire extinguisher

For each team of three students:
- 1 convection box with candle
- 6 wooden splints
- 3 pairs of goggles

Procedure: Part A—The Social Skill

1. Discuss what it means to be polite when doing team work.
2. Discuss the strategies your Unit 2 team used when practicing this skill.

   Share what seemed to work and what didn't work.

3. Record three ways your teammates can be polite to each other.

Procedure: Part B—The Box

1. Prepare your work space for the safe use of the convection boxes.

> ▲ **CAUTION:** Safety procedures include moving all papers and extra notebooks to the side of the classroom, tying back long hair, wearing goggles, and moving slowly while the boxes are in use. **NEVER LEAVE THE BOX UNATTENDED.**

2. Stand the box on its side, with the tubes up in the air. Take off the lid of the box.

   This is the Tracker's job.

3. Carefully light the candle and place it directly under one of the tubes (see Figure 10.1).

   This is the Manager's job. Be careful not to place the candle too near the back or sides of the box.

4. Put the lid on the box.

5. Carefully light a wooden splint and then blow it out. Lower the smoking splint 1 or 2 cm into the tube above the candle.

   This is the Communicator's job.

Figure 10.1

Before you light the candle that goes in the convection box, be certain that you've taken all the precautions you can to avoid fire hazards.

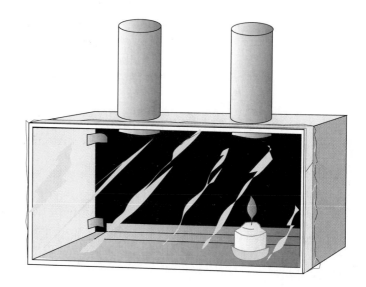

> ▲ **CAUTION:** **A smoking splint can still cause burns and fires because it is very hot. Be careful not to touch yourself, a teammate, or the box with the smoking part of the splint.**

6. Observe what happens to the smoke.
7. Lower the smoking splint 1 or 2 cm into the other tube and observe what happens.

   Take turns.

   Notebook entry: Record your observations.
8. Extinguish the candle.

   This is the Tracker's job.
9. Return the materials.

Wrap Up

On the basis of your experience with a convection box, answer the following questions as a team. Record your answers in your notebook. Be sure each of you can justify your answers in a class discussion.

1. How did the smoke help you observe what the air was doing?
2. What do we call moving air?
3. If you could stand inside the convection box, where would you be standing if you feel air sinking down on you?
4. Where would you have to stand inside the box if you wanted to feel air rising upward?
5. Would you find rising air above a warm area of the earth's surface or above a cool area?

6. Rate your team on politeness: good, fair, or poor. If you rated yourself a fair or a poor, modify or add to the three ways your team decided to be polite.

READING:
Changing Weather Patterns

If you've ever lived in a place where the temperatures can plummet from comfortably warm to freezing cold in just a few hours, you know how drastic weather changes can be. If you ever have been soaked in a rainstorm, trapped in a snowstorm, or burned by too much sunshine, you have experienced the results of weather patterns. To understand why certain weather patterns exist, think about some of the things you've experienced in this chapter.

Stop and Discuss

1. In what places have you seen water during this chapter?
2. Explain how the water got to those places.
3. Think of a method to test your explanation.

The air around you has water in it, water that you cannot see under normal circumstances. Water in the air is present in microscopic particles. If you live where it is very humid during the summer, you probably have experienced the feeling of water in the air. If you live in a dry area, you may have never even detected the water that's in the air and all around you. When the water particles in the air come in contact with something cool, the tiny water particles in the air clump together to form drops of water. This process is called **condensation.**

Stop and Discuss

4. What happens when you leave a glass of water out for a day or two?
5. What would happen if you left the glass out for a week or more?

When tiny water particles move into the air, the process is called **evaporation.** People often think that when water evaporates, it disappears, but in fact the small particles of water move into the air.

This process of water evaporating and then condensing happens all the time in nature. This is because heat from the sun is constantly warming the earth's surface. Warm water then evaporates from lakes, rivers, oceans, puddles, and moist ground. As the moist air rises away from the warm surface of the earth, the air cools. Soon the air is so cool that the water in it condenses into tiny water droplets and forms clouds (see Figure 10.2).

Explore ■ *Explain*

Figure 10.2

Because water is always losing heat energy (and condensing) or gaining heat energy (and evaporating), water particles are constantly moving.

As the air cools, water condenses and forms clouds.

When a lot of water condenses, the droplets in the clouds become larger and heavy enough so that they eventually fall as rain.

Stop and Discuss

6. Think back to the fish bowl demonstration. How could you keep the cycle going? See Figure 10.3 if you need a reminder.

Figure 10.3

When your teacher originally set up the fish bowl with the water in it, the sides of the fish bowl were dry. Where did the water droplets come from?

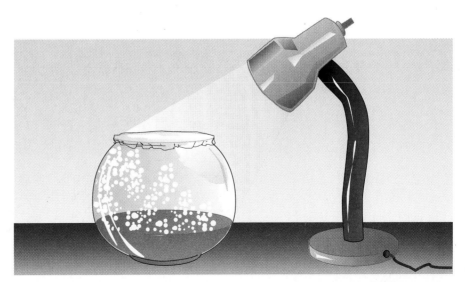

Explain

The pattern of water movement on the earth is called the water cycle. Water is constantly on the move. In some places it is going into the air, or evaporating, and in other places it is coming out of the air, or condensing.

The movement of water is a major part of the weather. As you think about the way water moves on the earth, remember that when rain or snow is falling, water is moving out of the air. In another place, where perhaps it is a sunny day, water is evaporating and moving into the air.

Stop and Discuss

7. Think about your explanations for why you saw water on the can and in the fish bowl. How could you change those explanations now?

8. Look at the diagram in Figure 10.4. The teakettle is plugged in, and the glass plate is cold. Describe what will happen as the water in the teakettle continues to heat. You may draw and label a diagram in your notebook or write several sentences to explain what will happen.

But *why* does water move? The short answer is because of heat. The sun's energy warms the earth's surface, including the water on the surface. If the water is heated enough, it changes from a liquid (the way we usually think of it) to a vapor, or a gas. When water is a gas, its particles are far apart. When it's a liquid, its particles are closer together. When water loses heat energy, it changes from a liquid to a solid. In other words, it becomes ice.

The heat energy that warms the water on the surface of the earth also warms the air. This causes the air to move. This moving air is the wind you feel outside. Wind is one of the factors behind weather patterns.

Figure 10.4

The water within this teakettle is being heated. What will happen to the water?

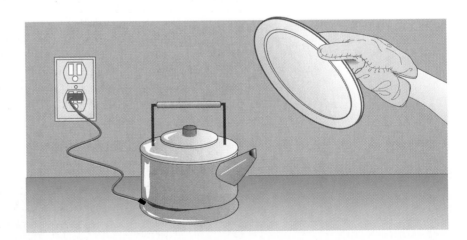

Explain

Figure 10.5

This diagram shows air particles in a convection box.

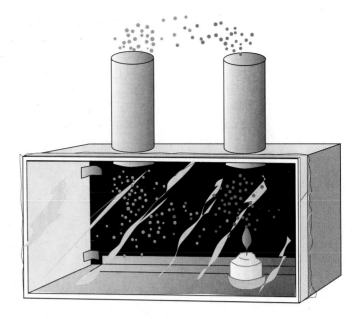

But how does heat energy cause wind? As the sun's energy reaches the earth's surface, the heat is absorbed unequally. This happens in part because the earth's surface is covered with land, as well as water, and solid materials absorb heat in a manner different from the way water does. Furthermore the sun's rays strike the earth more directly at the equator than they do at the poles. As a result the air in some places is warmer than in other places.

Stop and Discuss

9. Look at Figure 10.5. What parts of the convection box represent the following:
   - a warm place like the land at the equator
   - a cool place that does not receive direct rays from the sun
   - the wind
10. Describe the patterns you see in Figure 10.5.

The type of circulation or movement shown in Figure 10.5 is a **convection cell.** When air moves in a convection cell, you can feel it. For example, when cold winds in the Northern Hemisphere rush from the north to the south, those winds are part of a convection cell. In fact, one definition of **wind** is the air moving horizontally in a convection cell.

Stop and Discuss

11. As you can see in Figure 10.6, Al volunteered to become small enough to fit into a convection box. In your notebook describe what Al is experiencing with the air around him.

Figure 10.6

Al is standing exactly in the middle of the box, between the two tubes.

12. What would Al experience if he moved to his right? To his left? Forward? Backward?

13. If you have seen a hot air balloon like the one in Figure 10.7, you have seen an object floating in the air. How can this happen?

14. Why would an object sink?

15. In Unit 1 you placed pieces of wood, cork, and wax into several different liquids. Why would the same object float in one liquid but not in another?

Warm air inside

Figure 10.7

Why do you think hot air balloons are able to float?

Explain

Think about how objects can float on water. A piece of wood, for example, can float on water because the wood pushes down (we sometimes say it weighs) less than the water pushes it up.

When objects float, it is because the objects push down with less force than whatever is pushing them up. The force that allows something to float is called the **buoyant force.**

You have read about many factors causing weather patterns: the movement of water (evaporation and condensation), the difference in the amount of heat received and absorbed by parts of the earth, and the buoyant force. A scientist named George Hadley thought about these factors and thought that the earth's winds should blow from a cold area (the poles) toward a warm area (the equator). To explain this idea, he proposed that the earth was surrounded by two large convection cells. See Figure 10.8 for an example of his model.

Actual wind patterns on the earth, however, are more complex. Because the earth has land masses that absorb heat in a manner different from the way the oceans do, there are different types and levels of winds, such as surface winds and global winds (see Figure 10.9). For example, some near-surface winds blow over land and are very dry. Other surface winds blow over oceans and pick up water vapor. Some surface winds blow beside mountain ranges and through canyons. Within three adjacent states, for example, surface winds could be blowing in three different directions. High above the earth, however, global winds usually are undisturbed by local features on the ground, and the winds can flow in fairly constant directions.

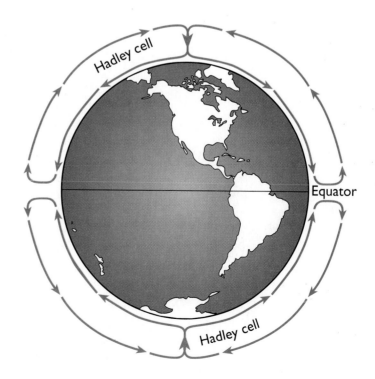

Figure 10.8

Hadley was the scientist who proposed this model. He envisioned two convection cells that covered the earth like bowls.

Figure 10.9

This diagram shows a low-level, or surface, wind blowing from north to south, and a high-level (global) wind blowing from the west.

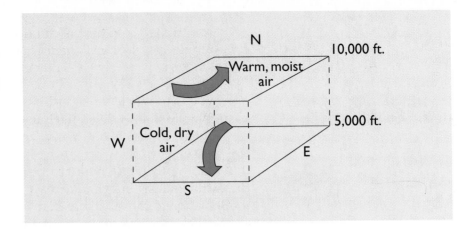

By now you know that wind patterns can be complex. You have read that water and land absorb heat differently. But still another factor affects wind patterns—the rotation of the earth. To see how the earth's rotation affects global wind patterns, complete the next investigation.

INVESTIGATION:
Winds above a Rotating Earth

You now have observed several patterns caused by moving air and water. In this investigation you will see whether you can apply your understanding to a new pattern. You will try to answer the question, How does the earth's rotation affect the movement of the wind?

Working Environment

Work cooperatively in your team of three. You will need a work space beside your desks or table in which each of you can stand and move freely. Use the roles of Manager, Communicator, and Tracker. As you work, use the strategies and ideas you developed for the social skill Treat others politely.

Materials

For each team of three students:

- 1 square of corrugated cardboard, at least 25-by-25 cm
- 1 pen or pencil
- 1 sheet of paper, 8½-by-11 in.
- masking tape, 5 cm
- 1 felt-tip pen
- 1 ruler, 30 cm (12 in.), or 1 piece of string, at least 30 cm long
- 1 pair of scissors

Procedure

1. Obtain the materials.
2. Trim the paper to fit your piece of cardboard.

Figure 10.10

This diagram shows the cardboard ready to spin on the pencil, the paper on top of the cardboard, and the ruler held horizontally above the cardboard.

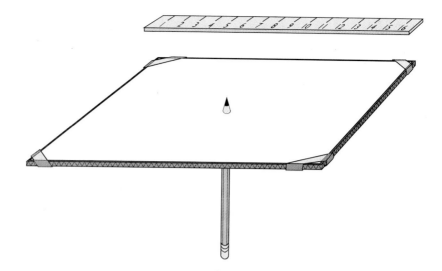

3. If there is no hole already, poke a pencil through the center of the cardboard as shown.

   The cardboard should spin freely on the pencil. See Figure 10.10.

4. Tape the sheet of paper onto the cardboard.

   The pencil should stick through the paper so that the paper and cardboard can spin together as shown in Figure 10.10.

5. Spin the cardboard on the pencil.

   The Communicator should do this.

6. Hold a ruler 5 cm above the middle of the cardboard.

   The Manager should do this as in Figure 10.10 (do not move the ruler).

7. On the spinning cardboard, draw a line on the paper with a felt-tip pen while keeping the pen against the ruler.

   The Tracker should do this as shown in Figure 10.11.

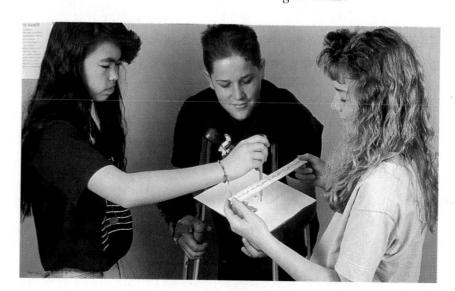

Figure 10.11

As the cardboard is spinning, move the pen along the ruler.

8. Take turns switching jobs so that each teammate has a chance to draw a line across the paper.

9. Read the Background Information.

Background Information

Coriolis Effect

If an object moves in a straight line across a spinning surface, the path the object leaves behind is a curved line. This is because the *surface* is spinning. When something moving in a straight line appears to curve, but in fact it is the surface beneath it that is curving, scientists describe this phenomenon as the **Coriolis effect.**

Scientists who observe wind patterns have noticed that the winds do not move in straight north and south lines across the surface of the earth. Scientists have noted that in the Northern Hemisphere, winds curve in a counterclockwise direction and that in the Southern Hemisphere, winds curve in a clockwise direction. The result is that winds tend to curve in more complicated patterns than we showed in Figure 10.8, the diagram of Hadley's convection cells. Instead wind patterns more closely resemble the diagram in Figure 10.12.

Figure 10.12

This diagram of the Coriolis effect shows that the winds do not move in straight lines from north to south. The arrows in this diagram indicate which way winds are blowing in convection cells. The darker portion of the arrow indicates the near-surface winds, and the lighter portion of the arrow indicates the high-level winds.

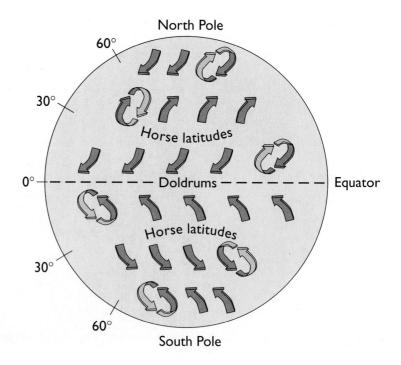

Winds appear to us to curve because the earth is rotating beneath them. Across the United States, winds tend to move from west to east, so most weather patterns also move across the continent from west to east. Figure 10.12 indicates the global wind circulation patterns, but remember that the local wind patterns you experience during a storm might be completely different.

Wrap Up

Conduct research to answer the following questions in your notebook.

1. If the ruler is straight and you moved a pen along it, why did you observe what you did?
2. What does *rotate* mean?
3. What evidence do you have that the earth is rotating?
4. When your area has rain or snow storms, in which direction do the storms tend to move?
5. Name one weather trend or cycle in your area.
6. How often is the weather forecast accurate? (If you don't know, watch, listen, or read about 5 days' worth of weather forecasts and compare them with what you actually observe.)

SIDELIGHT

In the Doldrums

Sailing ships once followed the routes of the prevailing winds across the globe to carry their cargoes from one continent to another. But some places on the globe do not have prevailing winds.

Have you ever heard anyone say that he or she is "in the doldrums"? What did that person mean?

Above the hot tropical regions of the world, the air warms and rises. On the surface of the land and sea, only light breezes blow. The near-surface breezes often change direction and never blow strongly in any one direction. This area along the equator was a dangerous area for the great sailing ships of past centuries and was known as the "doldrums."

There was so little wind that the ships could often be stuck in one area for weeks.

Farther north and south of the equator lie two other regions that posed hazards to sailing ships. In these subtropical areas, the air that has risen from the equator cools enough so that it sinks. This cool, dry air creates an area of fair weather. Most of the world's deserts are in these regions. Sailors called these belts of calm, high-pressure air around the world the "horse latitudes." Some people say that this is because horses sometimes died of thirst when the sailing ships they were on were stuck in calm waters. Because of the work of the early explorers, modern-day sailors have learned to avoid the horse latitudes and the doldrums.

Figure 10.13

This photograph was taken by a satellite and shows clouds over the equator.

CONNECTIONS:
Picture This

Look at the photographs in this activity and answer the accompanying questions.

1. What processes have occurred to form these clouds over the equator?

2. Look at Figure 10.13. Where is rain *not* falling over South America? Draw a diagram in your notebook and label it. Explain your answer.

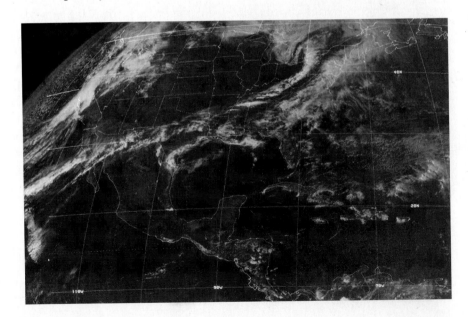

Figure 10.14

This satellite photograph shows clouds over the western part of North America.

Elaborate

3. In a convection cell, where would clouds tend to form?

4. Which direction will the clouds in the western United States move? (See Figure 10.14).

CONNECTIONS:
Science in Your Bathroom

You have just studied some of the earth's weather patterns and explored some ways that air and water move in cycles. Can you apply your understanding to everyday phenomena? Try to explain what happens in the following two situations you might encounter in your bathroom.

Situation A

You take a shower and the mirror fogs up.

1. Where does this water come from?

2. How does it get to the mirror?

Situation B

You take a hot shower, and the curtain billows in and brushes against your legs.

1. What causes the curtain to move like this?

2. What would happen if the water were cold?

Natural Events and Natural Disasters

Sometimes weather patterns do more than cause a light snow or a gentle rain. Some weather-related events in nature, as well as other kinds of events, cause problems for many people. As you begin the activities in this chapter, see whether you can decide what the problems are and how you might deal with them.

Figure 11.1

This photograph shows clouds of ash erupting from a volcano in Alaska on March 27, 1986. It has erupted explosively six times since 1812. It is similar to Mt. St. Helens, a volcano in Washington State. A major eruption of Mt. St. Helens occurred on May 18, 1980.

 CONNECTIONS:
When You See This, What Do You Think?

Sometimes life is hard. It's not just that the bus is late or that your locker doesn't open; it's the fact that somebody's entire house is washed away. To get ideas about the impact of some natural events, look at the photographs that follow. As you look at them, think about your own experiences with natural events that are out of the ordinary.

Look at each of the pictures in Figures 11.1 to 11.8 and read their descriptions. Then read the questions at the end of this activity. Think about how you would answer each question and discuss your answers with your teammates. Finally write answers in your notebook and be prepared for a class discussion of all the questions.

Figure 11.2

These cars were trapped in a lava flow from Mt. Kilauea, on the island of Hawaii. Lava flows from Mt. Kilauea stopped in 1960 and began again in 1983. Many different flows have occurred; the one that covered these cars occurred in 1987.

Figure 11.3

This building in the Marina District of San Francisco collapsed during the October 1989 earthquake. The building had three stories. The first and second stories collapsed; only the third story remains.

Figure 11.4

These buildings on the South Carolina coast were hit in 1989 by Hurricane Hugo's storm surge. As you can see, one building on stilts suffered less damage than the other one.

Figure 11.5

Heavy rains hit Montpelier, Vermont, early in 1992. Ice flows jammed a nearby river, and the water backed up. This caused immediate flooding, and traffic officers in Montpelier found themselves directing boats instead of cars.

Figure 11.6

This is Armour Station, Missouri, on October 16, 1953. The lakes at Armour Station vanished due to drought. The baked soil has cracks 25 to 35 centimeters (10 to 14 inches) deep.

The Bettmann Archive

Figure 11.7

The Yellowstone fires of 1988 created so much smoke that the sky appeared to be yellow.

Figure 11.8

Denver, Colorado, June 15, 1988. A tornado moves above the airport. Several tornadoes were sighted; two of them caused damage and minor injuries.

1. What evidence of natural events did you see in the photographs? Without looking at the pages, list all you can remember. Then look at the photographs again to see whether you missed any.

2. Rank the events you listed from most severe to least severe and justify your ranking.

3. What natural events have you heard about that have occurred in your community?

4. Often during natural events, people are more cooperative and helpful than they are at other times. Why do you think this could be true?

Smokey the Bear

In 1944 an advertising council began looking for a national symbol for fire prevention. They considered everything from a military theme to Bambi, the deer that survived a forest fire in a Walt Disney movie. Eventually they settled on a bear dressed in jeans and a forest ranger's hat. This bear was used on posters and in radio announcements and was very popular with the public.

Six years later a Forest Service helicopter pilot was circling over the remains of a fire in Lincoln National Forest in New Mexico when he spotted a bear cub. The cub was orphaned, badly burned, and weighed just over 3.5 kilograms (8 pounds). Veterinarians were able to treat his wounds, and when he regained his health, someone suggested that the bear be named Smokey and that he be moved to the National Zoological Park in Washington, D.C. Smokey became a popular animal at the zoo and remained a strong national symbol of fire prevention for over 35 years.

Smokey died in his sleep in 1976 and was replaced by Smokey II. Smokey II had been at the zoo since 1971. He had been found starving in the same forest where the first Smokey had been found. If you visit the National Zoological Park today, Smokey II is still there to remind us that "only you can prevent forest fires."

INVESTIGATION:
Miniature Events

Often you must experience something before you can understand it. But if a hurricane or a tornado strikes, you might not want to experience too much in person. Not, that is, if you want to be safe. As you move through the stations in this investigation, try to imagine how these miniature events are like the actual events.

Working Environment

Work cooperatively in your team of three. Use the roles of Communicator and Tracker. As you move from station to station, review your use of the skill *Stay with your group*, which you first practiced in Unit 2.

Materials

Station 1 (Fires)
- 1 drinking glass or jar
- 1 votive candle
- 1 box of safety matches

Station 2 (Droughts)
- 1 lamp
- 2 thermometers
- 2 pots or cups of soil
- 1 cardboard shade

Station 3 (Floods)
- 1 floodplain setup
- water supply and container
- water disposal container

Station 4 (Hurricanes)
- 2 empty aluminum cans
- 12 plastic soda straws

Station 5 (Tornadoes)
- 1 tornado bottle setup

Procedure: Part A—Stations

1. Move to your first assigned station.
2. Follow the instruction card for that station.
3. Observe what happens.

    Notebook entry: Record your observations and answer any questions that appear on the instruction card.

4. Move to other stations when your teacher says it is time.

    You should have notebook entries for all five stations when you have finished.

Wrap Up

In your notebook write an explanation for what you saw at each station and why it occurred. In each of your five explanations, include the following:

- station number,
- what the results you saw had to do with a particular type of event (fire, drought, flood, or so on), and
- what natural processes caused whatever you saw.

INVESTIGATION:
Presenting Events

You have survived a set of miniature events. By now you might be wondering what they had to do with the actual, potentially dangerous events. In this investigation your team's task will be to become experts on one type of event and to teach your classmates how the miniature events are like the real ones.

Materials

For each team of three students:
- 1 sheet of flip chart paper, newsprint, or 1 overhead transparency
- 1 colored marker or transparency marking pen
- any other materials you need for your presentation

Procedure

1. Read the information on your assigned topic.

    Your teacher will assign your team to read about fires, droughts, floods, tornadoes, or hurricanes. At the end of this investigation, your team will present this information to the class.

Working Environment

Work cooperatively in your same team of three. You will need a work space large enough for a piece of newsprint paper. One of you will be the Manager. Continue to practice the social skill *Treat others politely.* Also practice this skill when you meet in your combined team of six.

2. Discuss the reading with your teammates and make sure that you can answer these questions:
   a. What causes this type of event?
   b. How does it affect humans?
   c. How does it affect the environment?
   d. What do humans do about this type of event?
   e. What are some examples?
   f. How is the actual event like the miniature event you observed?

3. Meet with the other team that has your same reading.

4. Decide how you will divide up the presentation.

   One team of three should present answers to questions 2a, b, and c. The other team should present answers to questions 2d, e, and f.

5. Prepare and practice your presentation.

   For your presentation you may draw a poster or summarize your discoveries with words.

6. Give your presentation to the class.

   As you give your presentation and as you listen to others, show the rest of the class how politely your team can treat others.

Background Information

Jigsaw Reading 1: Fires

For a fire to start and continue burning, three elements are necessary: fuel, oxygen, and heat. The fuel source is anything that burns. In a forest fire, the fuel could be trees, bushes, grass, or debris on the ground. In a house fire, furniture, insulation, carpeting, clothing, and other household items serve as fuel for the fire. The hotter a fire gets, the more things will burn.

Oxygen is needed for something to burn. Because air contains oxygen, oxygen is usually available. The process of something igniting, or bursting into flames, is called **combustion.** During combustion oxygen combines with the fuel to make a fire. As the fire burns, it makes heat. This heat causes higher temperatures that help more things burn. For example, a wet log will not burn in a low-temperature fire. If the temperature rises, however, the water will boil out of the log and then the log will burn.

These three things—fuel, oxygen, and heat—are called the fire triangle. Just as a three-legged stool will not stand up if one of its legs is missing, a fire will not keep burning if one of these elements is missing. To put out a fire, fire fighters try to eliminate one or more of these elements. When fighting a forest fire, they often will remove the fuel from the path of the fire. They do this by creating fire lines—long ditches where they remove all the vegetation (plants and trees) by hand or by bulldozer. When the fire reaches the fire line, there is nothing left to burn. Fire fighters also spray water on a fire to reduce the temperature or cover it with dirt to cut off the oxygen supply, so the fire goes out.

While a fire is burning in a building or forest, convection currents are created that make the fire even harder to extinguish.

Stop and Think

Take a minute to review the convection box activity from Wind in a Box. As you review, think about these questions:

1. How did the smoke move in the box?
2. In a burning building or forest, what is the hotter air likely to do?
3. How will this pattern help the fire spread?

Weather conditions also play an important role in whether or not a fire keeps burning. On the one hand, dry, windy conditions increase the chances that a fire will start. On the other hand, rain and snow help keep fuels cool, so they do not ignite. Rain or snow also can smother the flames by keeping oxygen out. These wet conditions, however, only stop fires of low temperatures. If a fire is already burning well, the hot air rising in convection currents will evaporate the moisture before it reaches the ground. Or if the burning material on the ground is very hot, the rain and snow will boil away before the water can cool the fire. This is similar to what happens when you flick water onto a hot iron or pan—the water quickly sizzles away without cooling the iron or pan very much.

The temperature and moisture of the air can influence how easily a fire will start or how easily it will continue to burn. Temperatures below ⁻18° Celsius (0° Fahrenheit) make it very difficult for a fire to start. High humidity (lots of moisture in the air) also decreases the possibility of fire because the fuel is moist. Moisture keeps the temperature of fuel lower and reduces the chance of combustion.

Fires can be helpful in some places. For example, when fires burn in some forests, it becomes easier for new trees to grow. After a fire, seedlings receive more sunlight because older trees are not around to block the light. Some species of trees (lodgepole pines and jack pines) need the high temperatures fire provides. Heat allows their cones to open so that their seeds will be released.

Fires also can benefit wildlife and other animals that eat grass. When fires burn down trees or shrubs, sometimes grasses will spring up. People have begun to make use of this phenomenon in the plains region of western and central North America. By starting small, controlled test fires, people have found that grasses will grow back into areas taken over by shrubs. Grasses continue to grow as long as periodic burns occur.

Fires can start in three ways: by humans starting fires by accident, by humans starting fires on purpose, and by lightning. Fires that humans start, do not follow much of a pattern. In wilderness areas, however, patterns of fire occurrence are seen because trees, grass, and other plants act as fuel. These woodlands and grasslands can be ignited easily by lightning strikes whenever it is dry enough and there is enough fuel. When scientists examined the giant sequoias in California, for example, they found that fires occurred in the sequoia forests every 8 years. These fires did not damage the trees because of their thick bark. Instead the fires cleared the ground so that new trees could sprout. This pattern ended after humans started putting out fires in the parks where the sequoias grow.

People have affected the patterns of fire occurrence in other forests as well. One example is Yellowstone National Park. The summer of 1988 was hot and dry across much of the United States. The combination of heat, dry weather, strong winds, and several fires that had been started by lightning resulted in a forest fire that burned 1 million acres. It also started an ongoing controversy.

In 1988 the National Park Service had a policy called "let burn." This policy said that certain kinds of fires (primarily those started by lightning in isolated areas) should burn and follow their natural course. Then forests could go through their natural cycles of burning out debris and encouraging new growth. This policy had been in effect, however, only since 1970. From the 1930s to the 1970s, park officials had put out every fire. As a result, many areas of Yellowstone had at least 40 years' worth of accumulated debris on the ground. This debris was now available as fuel for the lightning fires of 1988.

The controversy about the 1988 fire centers on whether or not national parks, which people use for recreation, should be artificially maintained or be left more to natural processes. On the one hand, advocates of human control say that human needs, wants, and safety should come first, so we should put out and strictly control fires. On the other hand, advocates of natural processes say that natural burns help maintain balance in the forest. Without these fires, wild areas cannot regenerate, and the wildlife will lose valuable food sources.

Stop and Think

Would you want fire fighters to control fires in our national parks, or should we let natural processes control the fires?

Much more information is available on the Yellowstone fires. You may want to investigate further before you answer this question.

Melbourne, Australia, February 1983

Early in the 1980s, dry years in another location set the stage for fires. During 1982 and 1983, Australia had practically no rain at all. One careless spark could ignite the grasslands into a horrible blaze. In February 1983, a eucalyptus forest caught fire. In addition to dry conditions, eucalyptus trees contain a lot of oil that burns easily and causes explosions when the trees are hot. The wind carried the fire, and the small town of Maceddon was burned to the ground. By the time this fire was put out, 75 people had died, 8,000 were homeless, and 815,000 acres of forest and farmland had burned.

Fires also can occur in cities. On October 9, 1871, Chicago, Illinois, experienced a tragic fire. Chicago sits on the western shore of Lake Michigan. During much of the year, the difference in temperatures between the land and the water results in a constant breeze or wind blowing through the city. The year 1871 was very dry, and Chicago was a town with mostly wooden buildings and no fire department. Legend has it that Mrs. O'Leary's cow kicked over a lantern in the barn, and a raging fire resulted. The fire only burned for one day but left 100,000 people homeless and caused millions of dollars worth of damage.

When the citizens of Chicago rebuilt their town, they used stone for downtown buildings, laid the streets out on a logical grid, built steel skyscrapers, and established a fire department. These changes prepared the city to fight future fires.

Jigsaw Reading 2: Droughts

A drought (DROWT) happens when there is no rain for a long period of time. Droughts are slow-acting events, but they have affected more people than you may realize.

Some places on earth tend to have dry seasons, but enough rain or snow falls during the wet season to fill water-storage reservoirs. During the dry seasons, people can use the stored water to farm the land and to grow food. When the wet season occurs again, as expected, enough snow or rain falls to make up for long periods of dryness.

Stop and Think

Review the diagram of the water cycle from Chapter 10. Also think about weather patterns.

When water vapor in moist air condenses, clouds form and eventually rain (or snow) falls. Droughts occur when winds push clouds away from an area and so no rain falls there.

When a drought occurs, it may be because the wind never pushed the clouds over an area that needed the rain. For example, during the summer of 1988, winds did not push clouds over Minnesota and Iowa as they usually do. Instead the winds blew the clouds farther north, over Canada. That summer was wetter than usual in Canada and much drier than usual in states such as Minnesota and Iowa. So while there are drought years in some places, there may be wet years in other places.

A drought is a slow disaster. Many people can die in a severe drought because there is not enough water to grow food, and people simply starve. Other people become more susceptible to disease, and they die from sickness. Droughts have killed many people throughout history. In China one drought lasted from 1876 to 1879. It is estimated that between 9 and 13 *million* people died. People in other parts of China tried to send food, but some people were too weak even to come to the road and get food. It was hard for the wagons to get to the center of the drought-stricken area because the roads were poor. Sometimes the roads were blocked because thousands of people trying to leave the drought-stricken area had died on the road.

Similar disasters have happened in other places. Near the end of the 1800s, over 5 million people died in a drought in India. In 1921 and 1922, roughly 5 million people died in a drought in Russia.

More recently, droughts in Africa have claimed hundreds of thousands of lives. From 1968 to 1975, a devastating drought affected the Sahel region of Africa. It ended the way of life for many nomadic people. The nomadic people who did reach the cities were exposed to horrible conditions of filth and disease caused by overcrowding. Many of the children who survived suffered from deformed bodies and mental retardation, simply because they had not had enough to eat when they were very young. The African country of Ethiopia is another example; it has been drought stricken from 1981 into the 1990s.

Droughts also have long-lasting effects on the environment. For example, as a result of the African droughts of the 1970s and 1980s, the Sahara Desert grew larger. When a drought strikes, plants and animals die. During the summer of 1988, people in the United States observed that some of the water wildlife was dying. As the ponds dried up, bacteria known as clostridia were concentrated in the remaining water. (Some kinds of clostridia can cause a sickness known as botulism in animals and plants. If these harmful clostridia accumulate in food, they can kill people and animals.) In this case the harmful clostridia accumulated in the water, and many of the ducks that drank the water died of botulism.

Droughts also can make an area more susceptible to fire. The remaining plants dry out so much that they can burn easily. So if a drought happens, fires are more likely, and there is less water to put out the fires.

Historically, people have had very few defenses against drought. When a severe drought struck the central United States during the 1930s, many people had no other solution but to leave the area. Many of them moved all the way from Oklahoma to California. During those years such states as Oklahoma and Nebraska were part of what was called the Dust Bowl.

Once the drought of the 1930s was over, people in the United States began to look for better ways to protect themselves from drought. People planted trees as windbreaks to slow down soil erosion; this would help keep the farmland fertile. People also built reservoirs to store water. In addition, people began to store food for the years when droughts might strike again.

Jigsaw Reading 3: Floods

Even though water may be all around you during a flood, it actually can be difficult to find safe drinking water. This shortage of safe drinking water happens because water lines break and sewage lines overflow making the water unfit for humans to drink.

As you saw in one of the photographs at the beginning of this chapter, when a flood occurs, water covers land that is usually dry. Floods can happen suddenly, or the water can rise gradually. Floods can last for hours or days.

Floods happen because of a combination of events. Usually, for a flood to occur, three things must happen: There must be heavy rain, the soil must be so completely soaked that it cannot hold any more water, and stream channels or rivers must be filled with more water than they can carry.

As you might have thought already, heavy rain or melting snow is the primary cause of flooding. But what causes rain? Rain is water falling from the air.

Stop and Think

How did water (rain) get into the air? If you don't recall the answer to this, review the Chapter 10 reading on weather patterns.

Even if a heavy rain does fall, flooding will not necessarily happen. If the soil and vegetation absorb a lot of the water, streams may not overflow. If a stream had a very low water level before the storm, the stream channel may not overflow. If it does overflow, this is the beginning of a flood.

Floods happen in a variety of places—everywhere from near the oceans to inland deserts. Sometimes heavy rains occur near the coast because the air is near a large body of water. Here it is easy for the air to gain and then suddenly lose large amounts of water as the clouds move inland. In these coastal areas, floods can occur during or after heavy thunderstorms or during the heavy rains that accompany hurricanes. When hurricanes move onto land, a surge of ocean water pushed by winds also can cause flooding.

For places that are farther inland, floods occur mostly along rivers. Most rivers have channels in which the water usually flows, but they also have flat areas alongside them. The flat area alongside a river is called a **floodplain.** This is the land where water from a flooded river will go first.

Floods also can occur away from oceans, lakes, or rivers. Sometimes a thunderstorm will occur over a desert area. There are few plants, and the soil cannot absorb all the water. The water comes rushing down the hillsides and into the valleys below. Places with dry stream beds suddenly fill with gushing water that can throw aside anything in its path. Because these floods occur so suddenly, they are known as flash floods.

Flash floods often take people by surprise. People have come out of buildings in such places as Las Vegas, Nevada, to find that a sudden storm had washed away their car. Also, in Rock Springs, Wyoming, in July of 1989, people saw cars and trucks being swept down the street. People who see such events from safety are lucky. Often people who are caught in flash floods try to drive to safety. They rarely succeed. If you are in a car, the best idea is to get out of the car and immediately walk, run, or climb to the highest ground in sight.

Floods have caused millions of deaths. In the United States, flooding is responsible for more deaths than fires, droughts, hurricanes, or tornadoes. Some deaths occur because people don't want to leave their homes or cars.

Two severe floods in the United States happened when thunderstorms occurred at the edge of mountain canyons. After one such thunderstorm in 1972 in Rapid City, South Dakota, 232 people died in a flood. In Colorado, the Big Thompson River flood of 1976 killed 139 people. These were tragedies, but they were mild, compared with some of the deadliest floods in history. For example, in 1887 the Hwang-ho River of China overflowed its banks and caused the deaths of 800,000 people.

Clearly floods are the most damaging when people are not expecting them. One way that people try to eliminate the danger from floods is by building reservoirs. When a heavy storm occurs, the reservoir can hold the extra water. This works, however, only if the rain falls over the reservoir or over the river upstream from the reservoir. If the storm is downstream from a dam that holds back the reservoir water, there still will be a danger of flooding. Expecting a dam to protect people from flooding is not always a good idea.

It is important to avoid constructing buildings and homes on floodplains. Floodplains are usually flat and sometimes scenic, so people often want to build there. For safety's sake, however, city planners are wise when they limit such building.

What do floods do besides hurt people and cause property damage? For one thing they can enrich the soil. Some rivers have mild floods every year. Rivers carry nutrients and fine dirt, and when a river floods, it leaves these nutrients behind. During floods, the rivers also wash excess salts out of the soils. Annual flooding makes the soils very fertile. Some of the good farmland along the Nile River and the Mississippi River was a result of flooding. But people have built dams along many rivers such as the Nile, and so the annual flooding in some of these locations has stopped. As a result, the soils also have been drained of their nutrients and do not produce the same quality of crops.

Figure 11.9

This photograph shows how hurricanes appear from above. In the center of the hurricane is the eye.

Jigsaw Reading 4: Hurricanes

Hurricanes are storms that from above appear to be giant masses of swirling clouds (see Figure 11.9). They are extremely large storms, often 544 kilometers (340 miles) or more in diameter. Strong winds that blow at least 75 miles per hour and push water ahead of them cause widespread flooding. People who never have lived near a coastal area might not know how strong hurricanes can be.

Hurricanes form when warm, moist air rises, condensation occurs, clouds form, and thunderstorms develop. But the storms that can produce hurricanes occur only over warm oceans. Hurricanes develop fairly near but not exactly at the equator. This is because the Coriolis effect has no effect at the equator. So hurricanes begin slightly north or south of the equator in warm, moist regions.

Storms over the tropical areas of the oceans begin when dry **air masses** (areas with air of similar temperatures) from land meet moist air masses over the ocean. Because of the Coriolis effect, these meeting air masses begin to circulate in a counterclockwise direction (in the Northern Hemisphere).

As thunderstorms occur, more warm, moist air is drawn from the ocean's surface. This moist air is more buoyant than the air around it, and it rises. As the moist air rises, it cools because the air

Figure 11.10

This diagram shows the side view of a hurricane. Bands of thunderstorms are around a calm area in the middle.

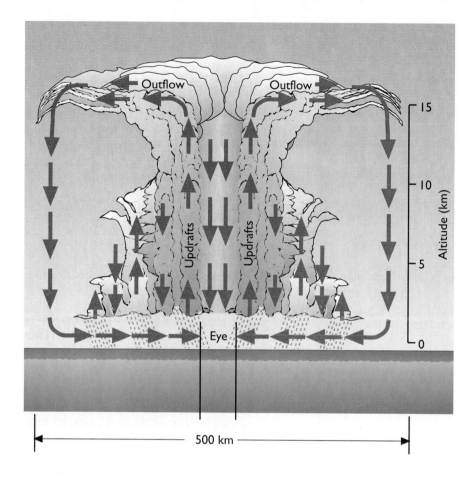

pressure is lower. As this happens, water vapor condenses, and eventually the cool air sinks. A thunderstorm begins, and within the storm many convection cells are set up.

Stop and Think

Review what you know about convection cells from Chapter 10. Then see whether you can find the convection cells in Figure 11.10.

Wherever warm, moist air is rising, ever larger quantities of air are drawn up into the storm. Updrafts continue. The clouds rise higher and higher. If the tropical storm has become a hurricane, it is a massive, swirling thunderstorm with very strong winds.

A unique feature of a hurricane is the **eye.** Out beyond the eye, warm air is spiraling upward. At the eye, which may be as large as 93 kilometers (58 miles) across, there are downdrafts where cool air is sinking. In the eye of the hurricane, skies are clear and the air is calm. People have learned that they must be cautious and not be fooled by the eye of the hurricane because the storm may appear to be over when it is not.

Figure 11.11

This diagram shows the location of Galveston, Texas.

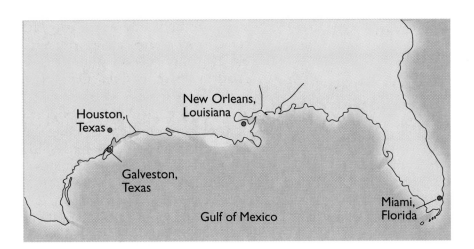

When is a hurricane really over? It depends. If a hurricane hits land, it no longer has the moisture it needs, and it eventually dies out, usually after doing much damage. A hurricane also can die out if it travels far enough north. There the air above the ocean is so cool that it stops rising.

One famous hurricane occurred in Galveston, Texas, in 1900. Galveston, Texas, is built on a large island off the coast of Texas (see Figure 11.11). People in Galveston knew that hurricanes had struck the island before, but they thought their homes were hurricane-proof. They were tragically wrong. When the 1900 hurricane struck the island, more than 7,000 people were killed. The bridges to the island were destroyed, and people could not escape even after they did realize that their homes were not safe. Many people drowned. One young woman was extremely fortunate. She clung to a tree branch that continued to float. When the floodwaters finally went down, she found herself 29 kilometers (18 miles) from home. Fortunately the only injuries she suffered were scrapes and bruises.

The floods that accompany a hurricane are brought on partly by a storm surge. A storm surge is a wall of ocean water that is pushed onto land by extremely strong winds. Storm surges can be over two stories high.

Sometimes people forget the lessons of earlier times. When Hurricane Camille struck the coast of Mississippi in 1969, no one was prepared for its strength. Camille was a category five hurricane, the strongest known. Homes and buildings along the coast were flooded and smashed. Even the leaves were stripped off the trees. People learned that the best way to be safe in a hurricane is to follow advice from the weather service. When you are advised to evacuate—do so!

Jigsaw Reading 5: Tornadoes

Tornadoes are swirling masses of air with wind speeds from 112 to 480 kilometers (70 to 300 miles) per hour. Their swirling winds can rip a building to shreds in a matter of seconds. Tornadoes strike suddenly and may accompany hurricanes. One hundred fifteen tornadoes occurred along with hurricane Beulah in 1967. Tornadoes are even more common in certain inland areas—they have ripped apart towns from Xenia, Ohio to Limon, Colorado. They can strike with deadly force.

The tornado that is thought to be the largest ever recorded occurred in 1925. It was a huge funnel cloud that bore down on Missouri, then moved on to Illinois, and then on to Indiana. It lasted far longer than tornadoes typically do and moved across three states, striking 23 towns. Along its deadly path, the tornado killed more than 600 people and injured hundreds of others. The tornado did more than $2 million worth of damage. This would probably be more than $1 billion worth of damage today.

Of course, there are also amazing stories of survival. In 1981 a man in Sumner, Texas, huddled in a bathtub during a sudden tornado. The tornado picked up the man and the bathtub and dropped them 0.4 kilometers (a quarter of a mile) from the demolished house.

But what causes tornadoes?

Stop and Think

To understand what causes tornadoes, review what you learned about winds in Chapter 10. Remember that air is constantly moving across the earth's surface. Near the equator, warm air is rising. Near the poles, cool air is sinking. But these **air masses** (areas with air of similar temperatures) do not sit still. They move, and sometimes they meet.

The following example describes how tornadoes can form over North America, even though tornadoes sometimes form over other continents too.

The air masses can move in the following pattern. Sometimes cool, dry air will move south down the eastern side of the Rocky Mountains. At the same time, warm, moist air will move north from the Gulf of Mexico, and the cold and warm air masses meet. As they do, the warm air is pushed up by the cold air and thunderstorms begin. The colder temperatures of high altitudes cause water vapor to condense from the formerly warm air. Warm, dry air also can move into the storm from the southwest. Severe thunderstorms can begin when dry air meets moist air, because the moist air will be pushed up by the dry air (see Figure 11.12).

Within the thunderstorm warm, moist air is rising and cool air is sinking. There are violent updrafts of air, and storm clouds build higher and higher. In addition to the north- and south-moving air masses and the upward motion of warm, moist air, winds may be blowing at higher levels. These winds from the west do at least two things. First, they push a thunderstorm toward the east. Second,

Figure 11.12

In this diagram a cold, dry mass of air that has moved south beside the Rocky Mountains meets a warm, moist air mass from the Gulf of Mexico. In this example, a warm, dry air mass from the southwest also enters the storm area.

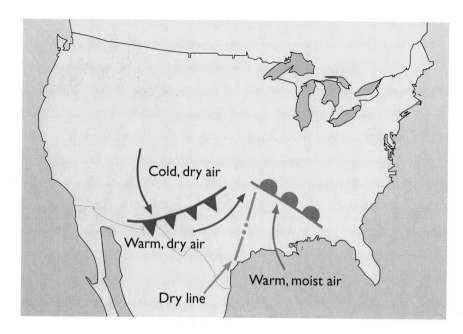

Cold, dry air

Warm, dry air

Dry line

Warm, moist air

they often are moving at such high speeds that they push air away from the top of the thunderstorm. So winds rush up from ground level, and the updrafts become even more violent (see Figures 11.13 and 11.14).

By this time you probably realize that winds are moving in several different directions. Winds from many directions and of different speeds are meeting within the thunderstorm, and eventually some of the violent updrafts begin to twist. When they do, the twisting cloud becomes a tighter and tighter swirl. If the

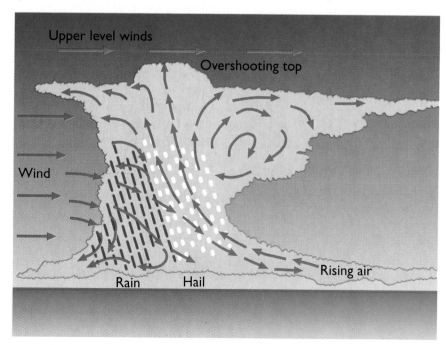

Upper level winds

Overshooting top

Wind

Rain Hail

Rising air

Figure 11.13

Side view: High-level winds push air away from the top of a thunderstorm, so other air rushes upward into the relatively emptier low-pressure area.

Figure 11.14

Winds are moving in several different directions. As they meet in the middle level of the thunderstorm, twisting of the updrafts begins.

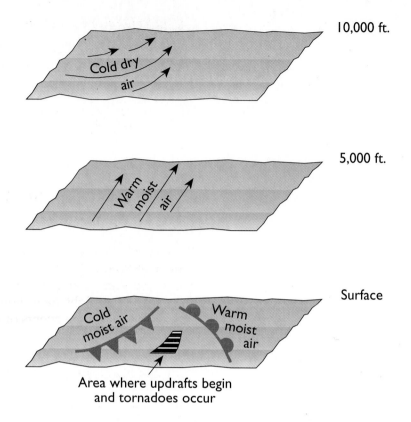

10,000 ft.

Cold dry air

5,000 ft.

Warm moist air

Surface

Cold moist air

Warm moist air

Area where updrafts begin and tornadoes occur

conditions are right, the swirling air mass will move down to the ground—this is a tornado.

What makes a tornado stop? The loss of warm, moist air. While the tornado has been forming, all the warm, moist air drawn up inside it has formed thunderclouds, which eventually lose their moisture as rain and hail. The rain and hail cool the ground around the tornado and the air near the ground. When the air near the ground becomes cool enough, it no longer rises because its weight (its buoyancy) is similar to that of the air around it. When there are no longer any air masses of contrasting temperatures, the tornado breaks down.

Wrap Up

After you have heard all of the presentations, consider the following questions. Write your answers in your notebook and prepare to discuss them with the rest of the class.

1. Go back to your wrap-up questions from the investigations Miniature Events. Write new answers based on the information you now have.

2. Rank the events according to a scale of your own choosing. (For example, you might use a scale from 1 = very dangerous to 10 = not dangerous to humans, or a scale from 1 = very rare event to 10 = very frequent event.)

3. Which natural events can make it easier for other events to occur?

4. Explain why the following sets of words are grouped together.

    a. convection cells: fires, floods, hurricanes, tornadoes

    b. water cycle: droughts, floods, hurricanes

    c. winds: droughts, fires

5. How did you benefit from working with a larger group on your presentation?

CONNECTIONS: You Decide

Think back to the events shown in the photographs in the first connections activity of the chapter. When people experience natural events of such strength, they often consider the events to be disasters. You might have your own ideas about whether or not these occurrences are events or disasters.

1. Read the following descriptions. Items *a* and *b* describe one event; *c* and *d* describe a second event, and *e* and *f* describe a third event. Identify what kind of event occurred and decide

whether or not you think each event was a disaster. Justify your answers in your notebook.

a. "Afterward there was more open space for flowers and grass. More animals are here. The views are beautiful."

b. "We lost at least $1 million worth of trees. We may lose our land altogether."

c. "The soil was washed clean of all the salt that had built up during the dry years. There's already a lot of new plant growth, and many more birds are nesting here."

d. "Our new apartment complex was completely destroyed. It will take months to wash out the mud, and even then, I'm not sure we would move back in."

e. "After the earthquake, water rose up along the fault, and a beautiful new lake formed. It's amazing, because the area around here always can use more water."

f. "Every building in town was knocked over. Seven people died. It was a real catastrophe."

2. Read the following and record your answers in your notebook.

People have not always used scientific explanations to explain these natural events. Some legends propose that an angry evil spirit causes storms. For example, the Seneca Indians of the Iroquois Confederacy told stories about Jijogweh (jih JOH gway), an evil monster bird that caused horrible storms and killed humans. In a Blackfoot Indian legend, the reckless horse Red Wind lived in the sky and caused tornadoes. When Red Wind decided to cause a little trouble, he came to earth, and his dancing caused the tornadoes. Why might people think that *angry* spirits caused natural events?

SIDELIGHT

Fire in London

In the 17th century, London was a town with very narrow streets and many houses and other buildings built close together. The houses had thatched roofs (woven straw). At this time houses did not have running water and the town had no fire department. The summer of 1666 was very hot and dry. In early September a baker's wood pile caught on fire. When the baker noticed the fire, he tried to put it out, but he couldn't because everything was dry and there was a strong wind. Finally he gave up and escaped through the roof.

The wind carried the sparks to other thatched roofs, and the fire spread quickly. By six o'clock the next morning, all of the houses and shops on Pudding Lane, where the bakery stood, were on fire. Most of the buildings on the next street over, Fish Street, and the houses on the London Bridge were burning too. The fire was out of control. People did not stay to fight the fire. Instead they collected their belongings and left for the countryside.

The wind blew nonstop for 3 days, and the fire kept spreading. Finally, 4 days after the fire had started, the wind died down and so did the large flames. The fire burned itself out completely, but not before 80 percent of the buildings in London were destroyed. Few people died, but 13,900 homes were burned, leaving three-fourths of the town's population homeless. The fire had burned over 400 acres, and 87 churches and cathedrals were ruined.

Explain ■ *Elaborate*

INVESTIGATION:
Twisters and Numbers

Have you ever experienced a tornado? Probably most people in your classroom have not—but some may have. Some people live where they are likely to see a tornado. It might seem that tornadoes just "appear" out of a cloudy sky, but there are patterns to when and where tornadoes occur. Because scientists have observed the weather patterns associated with tornadoes, fewer people die today during tornadoes than in previous times. In this investigation you will look at information about tornado occurrences. See what patterns you can find.

Working Environment

Work cooperatively in your teams of three. As you work Encourage each other to participate. Use the roles of Manager and Communicator. Sit together in your threesome configuration and remember that you are all Team Members.

Materials

For each team of three students:
- 1 copy of a Tornado Data Sheet for at least one state
- 1 sheet of graph paper
- 1 calculator

For the entire class:
- 1 overhead projector
- 1 overhead transparency, United States Map with Boxes
- 4 transparency marking pens of different colors
- 1 United States map, wall-size

Procedure

1. Pick up the graph paper and a Tornado Data Sheet with your team's number on it.

   Teams will count off to get a number.

2. Find your assigned state's location on a United States map.

   Your Tornado Data Sheet is based on information from a particular state. Be ready to locate this state on a map.

3. To help yourself remember your assigned state, sketch its shape and write its geographic location within the United States: north, south, east, west, midwest, southeast, southwest, northeast, northwest, or central.

   STOP: Is everyone participating? If not, how can you encourage more participation?

4. Study your team's Tornado Data Sheet.

5. As a team, find as many patterns in the data as you can.

   Averaging some of the numbers will help. If you do not know how to average numbers, do How To #9, How to Average Numbers.

6. Describe any patterns you find.

   Notebook entry: Record your descriptions of patterns.

Wrap Up

Read through the first two wrap-up questions. Think about how you would answer them, and then discuss the answers with your teammates. Record your answer to the first two wrap-up questions, and then prepare to share your answers with the class. Continue to work as a team on the third wrap-up question.

1. How many tornadoes usually occur each year in your assigned state? How many tornadoes usually occur each month?

2. Is there a month or a season when most tornadoes occur in your assigned state or in the surrounding region?

3. You are an insurance agent working for a company that writes policies all over the country. People buy policies to pay for damage that might happen to their homes. If disasters occur, your company pays. Predict where most disasters would occur. With your teammates, choose three states. Each of you should write a policy for one state. Explain what you would charge for home insurance in three different states you have heard about and why. For the three policies you write, you must answer the following questions:

    a. How are tornado patterns different or similar in the three states?

    b. Of all tornadoes that occur, where do most of them take place?

    c. When do most tornadoes occur?

    Encourage each other to participate by making sure that everyone has equal input in writing your policies.

Dust in the Wind

Have you ever traveled in a desert? Even if you haven't, you probably know that a desert has a lot of dry ground and few trees. When the wind moves across the desert, it picks up sand and dust and carries it along, up to a meter above the ground.

Maybe you have seen pictures of winds blowing dust and sand over the land. Sometimes tiny "twisters" form in those dust storms; these are called dust devils. If you've ever encountered one of these, you probably will remember it. A dust devil is a storm that develops from very warm air, particularly over bare ground. The warm air carrying sand and dirt is pushed up into the cooler air above it, and the air currents begin to swirl. The larger the area of bare ground, the larger the area of very warm air, and the longer the storm can last. This is why dust devils are common in deserts.

Although dust devils may look like tornadoes, they are not. A dust devil starts at the ground surface and moves upward quickly. A tornado begins in the clouds and moves downward quickly and violently.

People often think of dust devils as part of the American West. But they occur in other deserts too. In the Sahara Desert of Africa, dust devils have been strong enough to blow people out of tents.

INVESTIGATION:
Really Stormy Weather

As you know by now, hurricanes can be extremely powerful storms. But again, some patterns will allow you to make predictions. Imagine that you want to make predictions about hurricanes. If you look at some information that has been gathered already, you may be able to recognize certain patterns. As you look at this information, what patterns do you think you will find?

Working Environment

You will be working individually.

Materials

For each student:
- 1 sheet of graph paper
- 4 strips of transparent tape, 5 cm each

For the entire class:
- 1 world map, wall-size

Procedure

1. Obtain the materials.
2. Read Background Information, Part 1.
   This information follows the procedure.
3. Study the hurricane data chart. (See Figure 11.15.)
4. Decide how you can graph what the data chart is showing.
   If you need help, you might consult How To #3, How to Plot Data on a Graph.
5. Make a graph of the data and tape it into your notebook.
6. Read about the hurricanes assigned to you.
   These are in Background Information, Part 2.
7. Look at the description of
   - the hurricane's size,
   - how long it lasted,
   - where it went,
   - whether or not it hurt anyone, and
   - how fast its winds blew.
   Do this for each hurricane reading you were assigned.
8. Share your hurricane information with the class and discuss with your classmates how many groups of hurricanes there ought to be on the basis of the ones you read about.
   Try two or three groups and see which hurricanes fit into each.

Figure 11.15

This hurricane data chart lists the tropical storms and hurricanes from three different years.

Hurricane Data Chart

1987

Type	Name	Date
T	Unnamed	Aug 9-17
H	Arlene	Aug 10-23
T	Bret	Aug 18-24
T	Cindy	Sept 5-10
T	Dennis	Sept 8-20
H	Emily	Sept 20-26
H	Floyd	Oct 9-13

1988

Type	Name	Date
T	Alberto	Aug 6-8
T	Beryl	Aug 8-10
T	Chris	Aug 21-29
H	Debby	Aug 31-Sept 5
T	Ernesto	Sept 3-5
H	Florence	Sept 7-11
T	Unnamed	Sept 7-10
H	Gilbert	Sept 8-19
H	Helene	Sept 19-30
T	Isaac	Sept 28-Oct 1
H	Joan	Oct 11-22
T	Keith	Nov 17-24

1989

Type	Name	Date
T	Allison	June 24-27
T	Barry	July 9-14
H	Chantal	July 30-Aug 3
H	Dean	July 31-Aug 8
H	Erin	Aug 18-27
H	Felix	Aug 26-Sept 9
H	Gabrielle	Aug 30-Sept 13
H	Hugo	Sept 10-22
T	Iris	Sept 16-21
H	Jerry	Oct 12-16
T	Karen	Nov 28-Dec 4

Source: National Oceanic and Atmospheric Administration (NOAA).

Elaborate ■ *Evaluate*

Background Information

Part 1—General Information

Hurricanes begin as storms near the equator. These storms are sometimes called tropical storms. If the tropical storms' wind speeds become strong enough, they are then classified as hurricanes.

In the hurricane data chart that you will use in this investigation, some storms are listed as "T" (for tropical storm), while others were strong enough to be listed as "H" (for hurricane).

Part 2—Hurricane Descriptions

1989

Hurricane Gabrielle

Gabrielle moved off the coast of Africa on August 1. It was first classified as Tropical Storm Gabrielle. As it moved across the ocean, it was classified as a hurricane. Gabrielle turned north on September 4. It missed the Caribbean islands by about 480 kilometers and slowly stopped about 280 kilometers southeast of Cape Cod, Massachusetts. By September 12 it had lost most of its strength.

Gabrielle was a large hurricane. The eye was never less than 37 kilometers across, and sometimes it was as large as 93 kilometers (58 miles). Gabrielle's winds also were extremely strong, up to 112 kilometers (69 miles) per hour even 232 kilometers (144 miles) from the center of the storm. Gabrielle's winds generated ocean swells as high as 9 meters (30 feet). These swells hit shorelines from Bermuda all the way to Canada. The ocean swells caused eight deaths in the mid-Atlantic and New England regions. In some cases boats capsized while trying to enter or exit coastal inlets, and in other cases people were washed from jetties while watching the large swells.

1989

Hurricane Hugo

Hugo was first identified as a group of thunderstorms off the coast of Africa on September 9. This group of storms moved gradually westward and continued to gather strength. When wind speeds were measured on September 15, the best estimate for surface speeds was 260 kilometers (161 miles) per hour in the fastest moving parts of the storm. Hugo moved over the island of Puerto Rico and then struck South Carolina. Storm tides ranged from 2.4 to 6 meters (20 feet), and the storm surge was as high as 1.2 meters above normal, even several hundred kilometers up the coast. Heavy rains fell as far away as eastern Ohio. By September 22 Hugo had weakened to a tropical storm. It continued to move northward. On September 23 it was no longer classified as a tropical storm. It moved across eastern Canada and into the far northern Atlantic Ocean. Along the way, Hugo caused $10 billion worth of damage and claimed 49 lives (see Figure 11.16).

1988

Hurricane Gilbert

Gilbert began as a storm off the coast of Africa. It moved away from the African coast on September 3 and reached hurricane intensity on September 9. Gilbert moved south of Puerto Rico and crossed directly over Jamaica. Then Gilbert landed on the Yucatan Peninsula of Mexico. It continued to move northwest, crossed over the Gulf of Mexico, and made its final landfall in northern Mexico on September 16. At this landfall Gilbert's winds were 296 kilometers (183 miles) per hour—the strongest that had been recorded during any hurricane since 1969. Gilbert then moved north across Texas and into Oklahoma. Storm tides caused by Gilbert were as high as 4.5 meters. An estimated 318 people died as a result of Hurricane Gilbert, and $5 billion worth of damage was done.

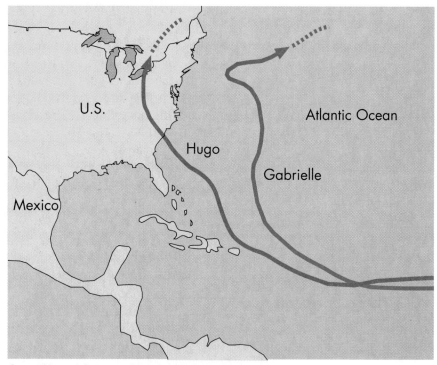

Source: National Oceanic and Atmospheric Administration (NOAA).

Figure 11.16

This map shows the tracks for two hurricanes from 1989.

Figure 11.17

This map shows the hurricane tracks for hurricanes Gilbert and Helene.

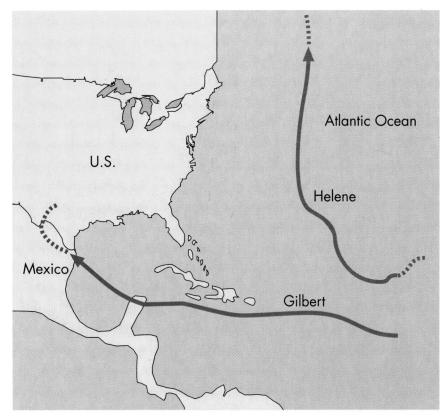

Source: National Oceanic and Atmospheric Administration (NOAA).

1988

Hurricane Helene

As Hurricane Gilbert was moving into the Gulf of Mexico, another storm was moving off the coast of Africa on September 15. By September 21 Helene was a hurricane moving through the midtropical Atlantic. On September 23 Helene moved northward. Helene's wind speeds were up to 232 kilometers (144 miles) per hour. Helene was a hurricane for 9 days, the longest-lasting storm of the season (see Figure 11.17).

1988

Hurricane Joan

Joan was a hurricane with a very unusual path. On October 11, 1988, Joan was classified as a tropical storm. It moved farther south than Atlantic hurricanes usually do and affected the north coasts of Venezuela and Columbia. Joan then moved back over the waters of the Gulf of Mexico. As the storm sat over the water, it strengthened. By the time Joan came ashore on the coast of Nicaragua, it was an extremely strong hurricane with wind speeds of 232 kilometers per

Figure 11.18

This map shows the hurricane tracks for Joan and Emily.

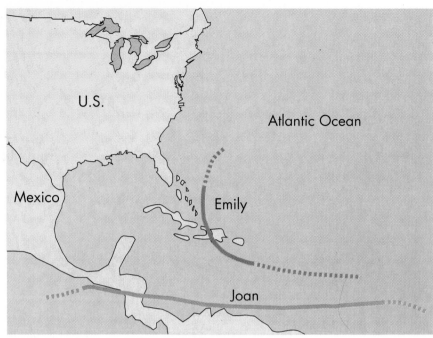

Source: National Oceanic and Atmospheric Administration (NOAA).

hour. Along its path, Joan was responsible for approximately 216 deaths, many in Nicaragua. Joan also was responsible for over $2 billion worth of damage. Once Joan crossed Central America, it continued as a storm into the Pacific.

1987

Hurricane Emily

Emily began as a storm off the coast of Africa and was classified as a hurricane on September 20, 1987. Emily was a small storm during this relatively mild hurricane season. Emily moved across the Atlantic Ocean. By the time the storm crossed over Bermuda, wind speeds had gusts up to 187 kilometers (116 miles) per hour. Emily was strong enough to pull out part of a dock, sending a cruise ship to float out into the stormy harbor. Luckily the captain and crew were able to keep the ship stable in the harbor as the hurricane passed over the rest of the island. Hurricane Emily caused approximately $35 million worth of damage to Bermuda, but fortunately no lives were lost (see Figure 11.18).

Wrap Up

Write answers to these questions in your notebook.

1. Of the hurricanes you read about,
   a. Where did most of them begin?
   b. Which direction(s) did they travel?
   c. What patterns are associated with hurricanes?

2. How might people use hurricane patterns to prevent loss of life and property?

Refer to the graph you made from the hurricane data chart.

3. Read the example below of a saying about a natural event and then

- make up one of your own or
- explain this one.

    In the Florida Keys, people have a saying:

    June—here soon

    July—stand by

    August—you must

    September—remember

    October—all over

CONNECTIONS:
People and Natural Events

Discuss the following questions with your classmates.

1. Which kind of natural event would you rank as the worst disaster and why?

2. How can people use patterns to help themselves cope with natural events?

Making Decisions and Solving Problems

People try to protect themselves from the dangers of natural events. For example, people try to reduce the hazards of fires and droughts by storing water. They protect themselves from tornadoes by having basements or storm cellars to wait in while a tornado passes by.

For protection against some natural events, people can even alter the designs of buildings. For example, this photograph shows one way that people protect their homes from high water: they've built houses on stilts. Now floodwaters that accompany a hurricane can pass beneath the houses. (In this example the people have covered the stilts with a lattice.) The people still may have to evacuate before the storm, but if the hurricane is not too severe, their houses on stilts may be undamaged by the storm. When people choose to live in any area, they usually have to adapt to the natural events in that place.

In this chapter you will use your understanding of natural events to make decisions. To learn about one situation in which people's decisions were very important, watch the video or film your teacher will show. See whether you can decide what problem people were trying to solve. What do you think of their solution?

INVESTIGATION:
Standing against the Wind

Wind can have destructive effects on the things that people build. In some parts of the world, people have to design buildings that can withstand extremely strong winds—winds blowing across a desert or perhaps out of a mountain canyon. Today people have found ways to strengthen buildings against the wind. Because of the materials builders use and because of the way architects design buildings, winds cannot damage most buildings very easily today. In this investigation see whether you can use materials and shapes to make a structure that will stand against the wind.

Working Environment

Work cooperatively in your team of three and use the roles of Manager and Communicator. You will need a work space that allows each of you equal access to a building project. As you Encourage others to participate, be sure to Treat others politely.

Materials

For the entire class:
- 1 model house
- 1 large box fan or blow dryer
- 1 wind scale, mounted on a box

For each team of three students:
- 1 pair of scissors
- 1 metric ruler
- 4 index cards, 3-by-5 in.
- 4 sheets of construction paper, 9-by-12 in. (1 for Part A, 3 for Part B)
- 1 sheet of white, unlined paper, 8½-by-11 in.
- 3 strips of transparent tape, 5-cm lengths (Part A)
- 3 strips of transparent tape, 5-cm lengths (Part B)

Procedure: Part A—Testing Materials

1. Watch the demonstration your teacher will do.
2. Think about how to build a stronger house than the one you just saw.
3. Read the Background Information.

   It follows the procedure.

4. Pick up the materials for Part A.
5. Cut out four rectangles of construction paper and four rectangles of white paper. Make them the same size as the index cards.

   Use one sheet of construction paper and one sheet of white paper to do this.

6. Tape four index cards together as shown. These are the walls of your index-card house.

   See Figure 12.1.

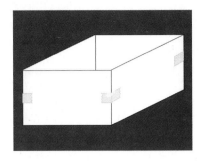

Figure 12.1

Tape your index cards together as shown to form a rectangular shape.

Figure 12.2

This diagram shows a triangular shape, which also is known as a wedge.

Figure 12.3

This diagram shows a tube shape, which also is known as a cylinder.

7. Tape the pieces of construction paper together in exactly the same way that you taped the index-card house together.

   This house should look just like the index-card house but be made of construction paper.

8. Make a house out of plain paper that looks exactly like the other two houses.

9. Take turns with other teams and test your team's three houses in front of the wind source.

   Your teacher will show you the wind source.

10. With your teammates, summarize the results for Part A.

    Notebook entry: Record your summary and describe which material stood up the best against the wind.

Procedure: Part B—Testing Shapes

1. As a team decide how to conduct a fair test to determine what *shape* of house will best withstand a strong wind.

   You should test at least the following shapes:

   - *a rectangular or cube shape,*
   - *a triangular shape, and*
   - *a cylindrical shape.*

   Use the following guidelines.

   - Each shape should be made of construction paper.
   - Each shape should use no more than one piece of construction paper.
   - Each shape should be closed on the top and sides, as shown in Figures 12.2 and 12.3.
   - Each shape should be at least 6 cm high and 6 cm wide.

2. After your teacher approves your plan, build the shapes.

3. Conduct your test of each structure's ability to withstand the wind.

4. Clean up the materials.

Background Information

A Fair Test

During this investigation you will be given instructions on how to set up a fair test. If you're not familiar with the idea of a fair test, you might think about a controlled experiment. Controlled experiments are basically the same things as fair tests. During Chapter 2 of Unit 1, you conducted a controlled experiment when you planted seeds and made sure that the conditions were the same for all the bean seeds, except for one factor, such as the amount of water the plant received. A fair test or controlled experiment means that *everything* is the same, except for *one* thing that you are comparing or testing. If this is not clear to you yet, read this description again before you do the wrap-up questions.

Wrap Up

Write answers to the following in your notebook.

1. Why did the three houses you built in Part A have to look exactly alike? Why did they have to be placed the same distance from the fan?

2. Which material worked best for building a wind-resistant house, and why?

3. Why was the entire class required to use construction paper for Part B?

4. Would the results of your test be different if you used materials other than construction paper? Why or why not?

5. If you were going to build a house that needed to withstand strong winds, in what shape would you build it? Why?

6. Describe how much your team has improved in treating others politely since the beginning of this unit.

INVESTIGATION:
The House on the Windy Plain

You now know a little about what types of houses might best withstand strong winds. You could go on testing factors such as materials and shapes for a long time, but maybe it's time to try to build the ideal, wind-resistant house.

Materials

For the entire class:
- 1 box fan or blow dryer
- 1 wind scale, mounted on a box
- construction materials (that your teacher will sell in the class supply store)

For each team of three students:
- 1 empty cereal box
- 1 metric ruler
- 1 pair of scissors
- 20 plastic tokens

Working Environment

Work cooperatively in your team of three. Use the roles of Manager and Communicator. Practice your Unit 3 skill. You will need a large work space for a building project and a common testing area. During the testing Treat others politely.

Procedure

1. Read the Background Information.

   It follows the procedure.

2. Read the following guidelines for planning your house:

   a. It must have a floor, walls, and a roof.

Figure 12.4

You may purchase construction materials from the class supply store. Your plastic tokens are worth 10 cents each for these supplies.

Foundation Connecting Materials	Wall and Roof Materials	Fastener Materials
Clay: 40¢ per sphere (pea size)	Index cards: 50¢ per card	Glue: 60¢ per bottle
Toothpicks: 10¢ each	Construction paper: $1.00 per half-sheet	Tape: 10¢ per cm
	Plain paper: 60¢ per sheet	Weights: 20¢ each

b. It must be large enough for a 6-cm person to be able to stand up and lie down inside.

c. The house must stand on the cereal box, which represents the ground.

You may decide to poke holes into this ground, but you may not make the holes bigger than a pencil.

3. Discuss the optional materials with your teammates. Decide what you will buy.

You must build the house by using the materials you buy at the class supply store. See Figure 12.4 for a list of the materials and their prices.

4. Decide how to build a house that stands best against the strong wind.

Use what you know about materials, size, and shape.

5. Draw your house plan on paper.

Make sure that you have allowed for a roof, walls, and a floor and that your house will be large enough for a 6-cm tall person.

6. Show your house plan to your teacher.

This is the Communicator's job.

7. Build your house according to your team's plan.

8. When your house is completed, use a pen or pencil to outline where your house is sitting. This line will mark the foundation of your house.

9. Participate in the class test of the houses and use the scale shown in Figure 12.5 to rate your house (and other houses).

Scale for Rating Houses

You might want to use the following point system:

Points:

4 It withstands the wind (needs no repair); or
3 The roof comes off the house (needs slight repairs); or
2 The house moves off its foundation (repairable); or
1 The house completely falls apart (destroyed).

Figure 12.5

You may want to use a point system for rating your house and those of other teams.

10. As a team discuss the characteristics of the house that stood up best against the wind.

11. With your teammates redesign your house, using what you learned from the class test.

12. Test your house again.

Background Information

Houses

People build houses to provide shelter for themselves or others. In some parts of the world, a house provides shelter against the heat and glaring sun. In other places a house protects against snow, rain, and cold. In some places houses must be able to withstand at least a mild hurricane. People have developed different types of houses to adapt to different situations.

A house typically has several basic parts: a foundation, a frame, floors, walls, and a roof. Many houses are called frame houses because the builders construct a wooden frame from strong pieces of wood and then attach the walls and roof to this frame (see Figure 12.6). Other houses, such as some log cabins, do not have a frame because the logs are the only materials used to construct the walls.

If you were building a house in certain parts of the world, you would want a house that could stand against strong winds. You would want a strong roof and walls, and you would want to fasten them securely to the foundation. As you work through this investigation, look for the choices that will make your house best able to withstand the wind.

Wrap Up

Discuss with your teammates all of the characteristics of your team's final house design. If your teacher calls on you in a class discussion, each of you should be prepared to explain your team's design and how you encouraged each other to participate. Then complete the following activity as a team.

Figure 12.6

This photograph shows the roof, frame, and foundation of a house.

1. In your notebook draw a sketch of your team's final house and describe what you did to make it wind resistant.

2. Imagine that you are trying to sell this house; design an advertisement that highlights its wind-resistant characteristics.

READING:
A Process for Solving Problems

There are hundreds of examples of people recognizing and solving problems. If people need food, they probably will look for ways to get food or to grow food; if people need shelter, they generally can find it or make it. People are inventive; they look for answers to their problems. When people recognize a problem, such as strong winds that could damage or destroy their homes, they usually look for a solution. For example, if you have a house that falls over in a strong wind, you probably first will recognize that there is a problem and then try to do something about it.

The knowledge that people use to design and construct things to solve problems is called **technology.** People often think that technology means computers, light bulbs, or space shuttles. Those objects were developed through a process of problem solving involving technology, but technology is much more than fancy machines and computers. Even something like a paper clip or a piece of tape resulted from the process of technological problem solving. For example, the paper clip solves the problem of loose papers, and we can use tape to repair torn papers.

Stop and Discuss

Did you think that you were using technology as you built your tiny house?

People have been using the process of technology for thousands of years. But in recent years, people have used technology to develop more new inventions every year. Sometimes new inventions build on other inventions; that is, the invention of one object sometimes makes it easier to invent other objects. For example, people invented wheels for carts, glass windows for houses, and large engines for boats and trains. Then about 100 years ago, people built new, smaller engines and combined them with wheels and glass to make the first automobiles.

Often as one problem is solved, another arises. For example, after people had cars to get from one place to another quickly, they wanted cars that were also safe. So designers investigated ways to build better brakes, safety belts, and child safety seats. This is one example of people making use of technological problem solving to find solutions.

CONNECTIONS:
The Popcorn Cube

Now this is technology I like.

Do you remember the cubes that you looked at in Chapter 6? During the investigation Think Like a Cube, you saw a cube that had a number on each side, but you could not see the bottom. Think about that cube and review your notes.

Now here's the problem. Imagine that you want to use one of those cubes for a new purpose: holding buttered popcorn. In addition you want to be able to eat the popcorn while you watch a movie, and you want the popcorn to stay warm for half an hour.

How would you change the cube? Spend some time thinking about this question before you continue this connections. Be sure to record your solution in your notebook.

As you thought about the popcorn cube, you were engaged in the process of technological problem solving. You were not trying to explain what was on the bottom of the cube, as you did in Unit 2. In that Unit 2 investigation, you were involved in a scientific explanation. This time you have been thinking about changes that you would make in the cube in order to solve a problem. In other words you were engaged in technological problem solving. People engaged in science seek answers to questions about the natural world. People engaged in the technological process try to solve problems associated with humans and their environment.

READING:
Decisions Are Part of the Process

As you imagined the perfect cube for holding popcorn, did the cube suddenly appear before your eyes? Probably not. Did the walls of your wind-resistant house put themselves together into the strongest

Explain

shape? No. You did the thinking and the building, and along the way you made decisions. As you look back on it, you'll probably see that you and your teammates had several decisions to make.

Stop and Discuss

1. List some of the decisions you made as you built your wind-resistant house.

When people use the process of technological problem solving, the situations almost always require decisions. So you might be wondering how people make good decisions.

One way to make a decision is to consider all of the **benefits** and **costs.** The benefits are all of the positive things that could result from a decision. The costs are all of the negative things that could result from a decision. Some people call the potential costs of making a decision its risks. For example, suppose you decided to spend four of your tokens to buy toothpicks for your model house. The benefit of this decision would be that you could use the toothpicks to support the walls of your house. The cost would be that now you could not use your tokens to buy extra tape or a weight. So here's a strategy for making decisions: First consider a certain choice. Then make a list of the good things that will or could happen (benefits) and a separate list of the bad things that will or could happen (the costs) if you make that choice. If you can predict that more bad than good will happen, usually the choice is not a good idea.

Stop and Discuss

2. Provide an example of something you've done that has had both benefits and costs.

One cost may outweigh many benefits, or sometimes one benefit may outweigh many costs. For example, someone might dare you to do something dangerous and even offer you money. The costs of refusing to do this might be that people will laugh at you and that you won't get the money. But you might decide that being safe is wiser. So perhaps you decide not to take the dare just because the one benefit (safety) is more important than all the costs.

Here's another example of a type of decision you already may have made. You're working on a model house for a contest. You were supposed to finish it last night, but you didn't. The contest is this morning at school. To make matters worse, your bottle of glue has dried out. In a kitchen drawer, you find a tube of glue that someone in your family used to make some home repairs. It's really strong glue, and you know it could work. You know, however, that this is a potentially dangerous glue. The warning on the package says that you cannot easily wash off this glue. If you got it on your skin, it could make your fingers stick together so tightly that you

Figure 12.7

Does this chart list all of the possible benefits and costs?

COSTS AND BENEFITS OF USING THE STRONG GLUE

Costs	Benefits
– fingers could get stuck together	– would finish house
– might have to go to doctor	– could (maybe) win the contest

would have to go to a doctor to get your fingers apart. But you really want to finish the house, and you think that you could be careful.

Stop and Discuss

3. What benefits and costs could you add to this chart (Figure 12.7)?

4. What will you do, and why?

By now you probably have decided what you would have done. The decision you had to make was whether or not to use that particular tube of glue. As you thought about it, though, you might have realized that you had several additional options. For example, if you decided to use the glue in the tube, you could look for a pair of plastic or rubber gloves to wear so that you wouldn't get the glue on your skin. You also could search for another bottle of glue or call a friend to see whether you could use some of his or her glue at school.

When you are making a decision, you usually have several options. Therefore, when you are making decisions, get as much information as you can. Imagine what the results of your decision would be like and then list the benefits and costs. Finally, decide.

INVESTIGATION:
How Did This Thing Get Here?

In this investigation you and your team will examine an everyday object. As you look at the object, try to determine what decisions the designers made when they were creating it.

I wonder how many problems this mug has solved.

Working Environment

Work cooperatively in your team of three. Sit in a triangular configuration. Use the roles of Manager, Tracker, and Communicator, as well as Team Member. Practice reviewing the skill that your teacher has assigned.

Materials

For each team of three students:

- 1 object assigned by the teacher

Procedure

1. Collect the object assigned to your team.

2. Pass your object around to each person in the group.

3. Place the object in the center of the table.

4. Read each of the questions below, one at a time. Have each member of the team offer an answer to the question.

   Notebook entry: Write one answer to each question after your team has discussed it. Answer all the questions.

   a. What is the purpose of this object?

   b. Does the object solve any problems? If so, which ones?

   c. Why do you think the designer chose this material?

   d. What other materials could the designer have chosen?

e. How do you think the designer decided on the size of the object? Is size important in how the object works?

f. How much could you change the shape of the object and still have it do the same job?

g. How did the designer choose the color of the object?

h. Would the object serve the same function if it were another color?

i. How would you change the structure of your object so that a small child could use it?

j. Suppose NASA decided to use your object on the next shuttle mission. The object would need to work in outer space, where the force of gravity is minimal. How would you change the design of the object so that people could use it in outer space?

Wrap Up

After your team has answered the questions, write a story in your notebook about how the designer developed the object. Work by yourself. Use your imagination and creativity in your story. Be sure to include your team's answers to questions *a* through *j* in your story.

The Tracker will give you 20 minutes to finish writing. Then your teacher may ask you to share it with the class.

 **CONNECTIONS:**
Balanced Decisions

As you know by now, when you want to make a wise decision, you must often consider the costs and benefits. In this activity see whether you can make some balanced decisions.

Follow these steps for the first situation:

- read about the situation;
- draw two blank cost and benefit charts, one for choice *a* and one for choice *b;*
- fill in your cost and benefit charts;
- answer any questions in your notebook; and
- explain what your decision would be.

Repeat the steps above for the second situation.

Situation #1

You have been selected to serve on a town committee. The town has raised money to build a new apartment complex for senior citizens. The town has selected a builder, and a company representative tells you about the two options for completing the project. You are asked to choose between these two options:

a. You could build a one-story complex of apartments at a convenient and scenic area in town. People would have easy access to stores and clinics. But because this is a scenic area, you are limited to one-story buildings; anything higher would block the view. The design of the one-story complex would allow some people to have small porches and yards, but there would be room for only 10 apartments. Currently the town has a waiting list of more than 30 people who want to move into the complex.

b. You could build a taller building on a different lot. This building could have two or three floors of apartments and room for more people than the first site. There would be no porches or balconies. The three-story building would allow 25 of the people on the waiting list to move in. The building itself would not be in as attractive a part of town, and it is more than 17 miles from the first, more convenient site.

1. Which choice has the most benefits and the fewest costs?

2. Which benefits are most important, and why?

Situation #2

In another building situation, your committee has to make a decision about safety. An old office building is being restored. It is a beautiful building in the downtown area, with woodwork and designs from the early 1900s. But here are the safety concerns.

a. Currently the building has a stairwell but no elevator. People want to make the building accessible to the handicapped, but there are problems.

In order to put in the elevator, part of the building must be torn out to make room for the elevator shaft. But for fire safety, the owners must have a stairwell. This stairwell must be of a certain size to allow many people to leave the building at the same time. Here's the problem: If the elevator is put in, the current stairwell will be too small for fire safety regulations. So the builders have proposed that a second, iron staircase could be built outside the building. With the inside and the outside stairs, the building would meet the fire code and allow handicapped people access to all the floors. Some people have argued, however, that the outside staircase is not a good idea because the building could then be vandalized more easily.

b. The second choice would be to make a larger inside stairwell, in addition to the elevator. This plan calls for no outside stairway. This plan would allow handicapped people access to all floors, as well as provide a fire escape and ensure a safe building. There are opponents to this plan too, however. Constructing the larger inside stairwell will cost much more. If the owners raise the rents in the building to cover these costs, small businesses may not be able to afford the rent, and the entire project may never be completed.

3. For the second situation, which solution would you choose, and why?

You're Probably Right

Have you ever seen a lava flow covering a highway? If you are like most people, you probably haven't, except perhaps in this photograph. By the time you've reached this point in the unit, you've looked at photographs of many unusual natural events. You have learned how you can make predictions about these events by using patterns. You've also seen examples of how people can use technology to solve problems. In this chapter you'll learn more about how people can make decisions about hard-to-predict events.

Many of the events that you have seen in this unit seem spectacular because they are unusual. They happen, but they do not happen every day. Also some of these natural events are unlikely to occur at the same time. For example, a community probably would not be struck by an earthquake, a hurricane, and a drought all at once.

As people decide how to make their homes and communities as safe as possible, they usually have to decide which kinds of natural events would be most likely to occur and cause them trouble. These decisions must be made because most people cannot afford to build homes or buildings that are safe against every type of event. For example, if you lived near the Gulf of Mexico and you built a house on stilts to protect your belongings from floods during hurricanes, you would not be as concerned with making your house earthquake-proof. You would have decided that it was more important to take precautions against hurricanes. But if you lived in California, your decision probably would be different, because of the frequency and severity of earthquakes there.

CONNECTIONS:
Those Difficult Decisions

People base their decisions on what they think probably will happen. People often watch weather forecasts to help them make decisions. If you have planned to attend an outdoor baseball game today, but then learn that heavy rains are predicted for this afternoon, you might decide not to go. Sometimes numbers are included in the forecast. You might hear a weather forecaster say on the news, "Tonight's forecast calls for a 20 percent chance of rain." Is this forecaster predicting that it is going to rain or not going to rain? Listen for more examples of forecasts as you watch the video your teacher will show.

INVESTIGATION:
It's in the Bag

How do people decide what risks they will take? As you've read before, people look for patterns and make predictions. Often they identify costs and benefits. Sometimes they even use numbers to show how certain they are about their predictions. As you and your teammates go through the investigation, see whether you can discover how people use numbers in predictions.

Working Environment

Work cooperatively in your team of three. Use the roles of Tracker, Communicator, and Manager. Agree on a social skill from this year that your team needs to practice the most. Report the skill you chose to your teacher. Work in your threesome configuration at your desks or a lab table.

Materials

For each team of three students:
- 20 plastic pieces, 10 red and 10 white
- 1 paper bag for holding the pieces

For the entire class:
- an additional 200 plastic pieces, 100 red and 100 white
- 2 paper bags
- 1 large map of the United States

Procedure: Part A—The Demonstration and More

1. Participate in the demonstration your teacher will do.

   See whether you can predict what the chances are that snow will fall in Bozeman.

2. Read How To #10, How to Make Sense of Percentages.

   You will use percentages in this investigation.

Our forecast calls for a 20 percent chance of snow.

Procedure: Part B—Bozeman, Montana

1. Set up a data table in your notebook.

   See Figure 13.1. You will use a data table like this to record your results for Parts B, C, and D.

2. Obtain the materials.

Did snow fall in Bozeman?		
Day	Yes	No
1		
2		
3		
4		
5		
6		
7		
8		
9		
10		

Figure 13.1

Copy this data table into your notebook.

3. Put two white pieces and eight red pieces into the paper bag.

Let the Tracker do this.

4. Gently shake the bag to mix the 10 pieces.

Let the Tracker do this.

5. Read about the following situation in Bozeman, Montana.

Situation A

As Isaac has pointed out, there is a 20 percent chance of snow for Bozeman, Montana today. Looking at the long-range forecast, the situation should remain much the same for the next 10 days. So for the next 10 days, for each day there is a 20 percent chance of snow.

6. Hold the bag above your teammate's eye level. (Now you will find out what actually happened in Bozeman.)

The Tracker should do this.

7. Draw out one plastic piece.

The Communicator should do this without looking into the bag.

8. Observe whether or not it snowed.

- *If the Communicator draws out a white piece, then it snowed in Bozeman that day.*

- *If the Communicator draws out a red piece, then it did not snow in Bozeman that day.*

9. Record whether or not it snowed on Day 1.

The Manager should do this on his or her chart.

10. Return the colored piece to the bag so that there are again 10 pieces in it.

11. Shake the bag to mix the pieces.

12. Repeat steps 6 through 10 nine more times, one time for each day.

The Manager should record whether or not it snowed on Days 2 through 10.

13. Take the pieces out of the bag.

Procedure: Part C—Baton Rouge, Louisiana

1. Remove the pieces from your bag and replace them with five red pieces and five white pieces.

The Communicator should do this. There should be 10 pieces all together in your team's bag.

2. Read the forecast for Baton Rouge, Louisiana.

Situation B

As Marie has said, we might get a little rain. The forecasters say there is a 50 percent chance of rain in Baton Rouge for both today and tonight. Tomorrow and the next day have similar forecasts. Both tomorrow and the next day each have a 50 percent chance of rain and thunderstorms.

3. Make a data table for Situation B.

You will use this chart to record whether or not it rained in Baton Rouge for 3 days.

4. Draw out one plastic piece.

This time the Manager should hold the bag, and without looking, the Tracker should draw out one piece.

5. Record whether or not it rained in Baton Rouge on Day 1.

The Communicator should record this on his or her table.

6. Return the plastic piece to the bag.

7. Repeat steps 4 and 6 two more times, once for Day 2 and once for Day 3.

The Communicator should record the results for Days 2 and 3.

8. Fill in your own data chart for Parts B and C.

Be sure that you have your team's results recorded for Bozeman and Baton Rouge.

9. Participate in the class discussion.

Notebook entry. Listen to find out how your team's results differed from or were similar to other teams' results.

Procedure: Part D—Anytown

1. Read about Anytown.

   The utility company in Anytown, U.S.A., has built a nuclear power plant. It wanted to make sure the plant would be safe. Anytown has the slight possibility of an earthquake. The chances are 3 out of 100 that a major earthquake will occur in any year. If a major earthquake hits Anytown, there is a chance that the nuclear plant would be damaged or would release dangerous pollutants.

2. Make a data table for recording earthquake occurrences in Anytown.

   Notebook entry: Use a full page and make two columns, one labeled "yes" and the other labeled "no." Allow space on your table for 50 years.

3. Discuss with your classmates how you will represent the probability of an earthquake.

   In this case let red pieces indicate an earthquake.

4. Participate in the class drawing.

Each drawing will represent what happened in a particular year. Remember that after each drawing, you must place the plastic pieces back into the bag.

5. Put away the materials.

Wrap Up

Discuss the following questions with your teammates and write the answers in your notebook. Each of you should be able to explain your answers during a class discussion.

1. With a 20 percent chance of snow in Bozeman, how many days did it snow for your team and the two teams nearest you?

2. Suppose the forecast for Baton Rouge, Louisiana, includes a 60 percent chance of rain for each of the next three days.

a. Is it possible that the next three days in Baton Rouge will have no rain?

b. Is it possible that it could rain each of the next three days? Explain your answers.

3. You read about a place with a 3 percent risk of a major earthquake occurring that could trigger a catastrophe in a nuclear power plant.

a. Is an earthquake likely to occur?

b. Is the site safe enough for a nuclear power plant? Why or why not?

4. Use your own rating system to evaluate how well you used the skill you chose. Choose a reward from the class reward list if you think it is appropriate.

READING:
Definitely, Probably, Maybe

Statements of Probability

You might wish that the weather forecaster could just *tell* you whether or not it is going to rain or snow. Why do forecasters say that there is a 25 percent *chance* of rain? They say this because they do not know for certain. We can use some types of patterns to make very accurate predictions. But other patterns, such as weather patterns, aren't always as predictable. We don't always have enough information, and sometimes things don't follow definite patterns. So sometimes we use what information we have to say what *probably* will happen.

Here's an example. Suppose one of your friends goes to the store 3 days a week but doesn't always go on the same days. In some weeks he goes on Monday, Thursday, and Friday. Other weeks he goes on Tuesday, Saturday, and Sunday, and so on. Now suppose someone asked you whether your friend was going to go to the store on a Thursday 3 months from now. Could you give a definite answer? No, because you don't have enough information to make the prediction—your friend hasn't established an exact pattern of when he goes to the store. But there *is* a pattern. You know that your friend goes to the store 3 days out of every 7. That's a little less than 50 percent of the days. So although you couldn't state for sure whether or not your friend would go to the store on that Thursday, you *could* say something like, "There's a little less than a 50 percent chance that he will go to the store on that Thursday."

We call statements like this statements of probability. They're statements that say what could happen or what is likely to happen, but they don't say for sure what *will* happen. Think back to when you selected plastic pieces to determine whether it snowed in Bozeman, Montana. Each team used the same probability—a 20 percent chance of snow each day. But each team probably did not get the same results. Maybe it didn't snow at all for some teams, and other teams may have had snow 4 days or more. Of course, for most teams, there probably were more days of no snow. That's because a probability of 20 percent, or 20 out of 100, is a fairly small probability. If the probability of snow were 70 percent, you could expect that most teams would end up with more days of snow than they did with a probability of 20 percent.

Stop and Discuss

1. For Bozeman and Baton Rouge, why did you have to replace the plastic piece each time and always draw from a bag that had 10 pieces in it?

2. If the weather forecasters in Bozeman are predicting a 20 percent chance of snow, would you still plan a drive through 50 miles of the Montana countryside? Explain your answer.

Making Decisions Based on Probabilities

In Chapter 12 you learned that one way to make a decision is to weigh the costs and benefits that would result from the possible decisions. But sometimes you don't know what the costs and benefits will be. For example, if you're planning to go to an outdoor baseball game on Saturday and you know that it's going to rain, you can decide whether seeing the game (a benefit) is worth sitting in the rain (a cost). But people don't always know for sure whether or not it's going to rain. What if there's a 70 percent chance of rain? This makes your decision more difficult because you only know what is *likely* to happen.

Stop and Discuss

3. During the investigation It's in the Bag, it might have seemed to you that a 20 percent chance was a small risk for snow but that a 3 percent chance was a high risk for a major earthquake that could damage a nuclear power plant. Why might people think differently about probability for snow and for damage to a nuclear power plant?

4. Describe another pair of events in which one could be an inconvenience and the other could be a disaster.

5. Describe how the type of event (an inconvenience or a disaster) changes the percentage of chance that you are willing to accept.

INVESTIGATION:
What Are the Chances?

The sun *always* rises in the east. The winds *probably* will increase later today. A snowstorm in July is *unlikely* in Florida. There is a *small chance* that I will pass my history test without studying. These statements all have something in common. Each one describes the chances or probability of an event occurring. In this investigation you will have an opportunity to rate such statements according to how likely or unlikely you think they are to occur.

Materials

For each student:
- 1 copy of Words by Chance
- 1 pair of scissors
- glue or tape

Procedure

1. Draw a line like the one shown in Figure 13.2 near the top of a page in your notebook.

Working Environment

You will work by yourself.

Figure 13.2

Draw a line like this in your notebook.

Low Chance   High Chance

2. Write "Low Chance" on the left edge of the page and "High Chance" on the right edge.

   This will be your scale.

3. Cut out the words from Words by Chance.

4. Lay the pieces of cutout paper on your open notebook, below the line you have drawn.

5. Group the words so that ones describing approximately the same chance of something happening are together.

6. Then arrange these groups on the paper in order from low chance to high chance.

7. When you are satisfied with your arrangement of the cutout words, attach or copy the words into your notebook.

Wrap Up

1. Use the words from Words by Chance to describe the probabilities that the following events will occur.

   a. The sun will rise tomorrow.

   b. A coin will land head up.

   c. You will be absent from school at least 1 day this year.

   d. A blizzard will occur in Hawaii in August.

   e. The next baby born in your town's hospital will be a girl.

   f. An earthquake will occur in California during the next century.

2. Look at the probability scale in Figure 13.3 and draw one just like it in your notebook. Leave at least 10 centimeters of vertical space between this scale and the scale you drew in step 1.

   Answer the following questions, using percentages and enter the numbers above the line of the scale.

   a. On the probability scale you have drawn, what number should you assign to *certainty*?

   b. What number should you assign to *even chance*?

   c. What number should you assign to *no chance*?

   Answer these questions, using percentages.

3. What is the chance that the sun will rise tomorrow?

Figure 13.3

Note that the marks on this scale divide it into 10 parts.

No Chance   Low Probability   Even Chance   High Probability   Certainty

4. What is the chance that the next baby born in your town's hospital will be a girl?

5. Suppose that the last 15 babies born in the hospital were boys. What is the chance that the next baby born in your town's hospital will be a boy?

6. What is your chance of winning a drawing if you bought 5 of the 100 tickets sold?

7. What is your chance of winning if you bought 25 of the 100 tickets sold?

INVESTIGATION:
What Will Happen Here?

By now you might realize that you can use probability to make predictions about all kinds of things, from whether or not you should wear a rain jacket to school one day to whether or not a blizzard will occur. In this investigation you will use probability to decide which types of events are most likely to occur at a given location.

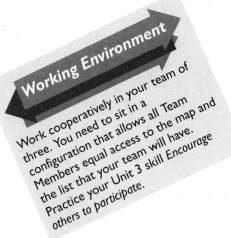

Working Environment

Work cooperatively in your team of three. You need to sit in a configuration that allows all Team Members equal access to the map and the list that your team will have. Practice your Unit 3 skill Encourage others to participate.

Materials

For the entire class:
 ■ 1 wall map of the United States
For each team of three students:
 ■ 1 copy of the Natural Events List
 ■ 1 copy of the U.S. Map with Location Dots

Procedure

1. Obtain the materials.
2. Read the Background Information.
3. Look at the map you have.

   Notice where the letters and dots indicate six locations.

4. Read the Natural Events List.
5. Decide which numbered set of events should go with which location.

   Mark a number beside each letter to indicate which set of events is likely to happen at each location.

   STOP: Remember to encourage others to participate.

6. Discuss your opinions with your teammates.

If you cannot agree, put your initials beside the number you wrote. Be ready to justify your answer.

7. Participate in the class discussion.

Listen as other teams explain why they think a particular set of events would occur at a certain location.

Background Information

During this investigation you will read about the probability of certain natural events occurring. As you try to decide which events would occur at each location, you might find that the following information is useful.

Hurricanes mostly affect the land only along coastlines. In the United States, hurricanes usually occur in the eastern and southeastern parts of the country.

■ Tornadoes may occur almost anywhere in the United States. Check your notebook and Chapter 10 to remember where most tornadoes occur.

■ Fires may occur in any location where there is fuel to burn.

■ Droughts may occur in any part of the country, although they usually are associated with the western, midwestern, and southwestern parts of the United States.

■ Snowstorms may occur almost anywhere in the United States, but they occur much more often in the northern and mountainous parts of the country.

Wrap Up

For the activities in this section, work individually and write answers in your notebook.

1. Choose one of the following questions to answer:

■ What is the best method you have found of encouraging others to participate?

■ What is one example you saw during class of someone during class effectively encouraging someone else to participate?

■ How did you encourage your teammates today?

2. Choose one of the six locations in the investigation. Write down which set of events your team decided would be most likely to happen there. Then write a story about what has happened at that location during the last year or the last 100 years or what will happen in the future.

As you can see, Isaac, Marie, and Al are satisfied that they have solved a problem with a model bridge they have built. If you think over what you know about houses and bridges, perhaps you can decide what problems the bridge solved.

1. What are some of the natural events that architects and engineers must consider when constructing buildings and bridges? If you don't remember, look back at the readings in Chapters 10 and 11.

Natural events have the power to destroy human lives and the structures that people live in. When such destruction occurs, we consider the event a disaster and make an effort to change the way we live or the way we construct buildings. People begin to adapt when they make observations about natural events and recognize the patterns associated with these events. Think back to Chapter 10 and review the answers you wrote in your notebook.

2. What are some of the patterns associated with hurricanes and tornadoes?

People have made use of patterns not only to predict events but also to predict how to solve problems as well. For example, when you designed houses and tested them in the wind, you might have noticed that some shapes and some materials always stood up well in the wind. As you made the observations, you found the key to solving a problem, and you learned more about making a house that would withstand a strong wind. With your new-found knowledge, you knew how to solve a technological problem.

3. Explain how the process of solving a technological problem is different from developing a scientific explanation.

An important part of technological problem solving is making decisions. Every technological decision involves costs and benefits. Sometimes the best answer or solution is not easy to find because one cost or one benefit may outweigh many of the other benefits or costs.

4. What is one example of a situation in which one benefit can outweigh many costs?

When we make decisions, we want to know what probably will happen. In some cases we use probability statements. People use probability statements when their decisions involve hard-to-predict events.

5. An example of a probability statement is, "There is a 40 percent chance of rain tonight." What does this statement mean?

So how do people use patterns to adapt to natural events? People look for patterns to discover what has happened, to predict what probably will happen, and to figure out how to solve problems that might result from the events. As you proceed through the next activity, see whether you can apply what you know about natural events, technology, decisions, and probability.

CONNECTIONS:
You Be the Judge

A citizens group has selected you to serve on a committee. A scenic area in a semi-arid location has been struck by a terrible drought. This is the third year of that drought. The committee must make some decisions that involve technology.

Solution A

One possible solution for the drought would be to pipe water to the area. A place not too far away has a reservoir with water in it, and this place is likely to have a rainy season this year. To make sure that there is enough water for both places, however, some of the people in charge of the reservoir have said that they would have to raise the height of the dam to allow the town to store more water.

Opponents of this idea say that if too much water is added to the reservoir, the weight of the added water will put too much pressure on the faults below and could trigger an earthquake. They believe that increasing the height of the dam is a dangerous idea. Several engineers have come to the area and have studied the problem. They have listed the probabilities of natural disasters continuing or occurring as follows:

- drought: 98 percent per year
- earthquake: 20 percent per year

Solution B

Other people have suggested that a solution might come from even farther away. Some areas along the coast have heavy rains and plenty of water. Large trucks could haul drinking water from there to the areas that really need it. Also the cost of hauling the water by truck will be much less expensive than raising the dam. Only if trucks hauled water for 6 years would the expenses be equal to the cost of raising the dam.

But some people have argued that these large trucks will cause extra air pollution and create a traffic hazard along narrow and scenic roads. Much of this coastal area is a forest wilderness that people are trying to preserve. They also feel that the noise of the trucks would create noise pollution and that the sight of so many trucks along the roads would be a type of visual pollution. This coastal area is a tourist area, and the increase in pollution may decrease the number of tourists and hurt the economy. Scientists have studied this option as well, and they have reported on the probabilities as follows:

- drought: 98 percent per year
- that air pollution will increase: 20 percent per year

1. Draw a chart that shows the costs and benefits for Solution A and another chart that shows the costs and benefits for Solution B.

2. Explain what you think should be done and why.

3. Describe at least one other possible solution to the problem. What are the costs and benefits of this solution?

U N I T 4

Patterns and People

So far this year, you have investigated patterns that people have discovered. People saw patterns in the changes of the moon and used those patterns to devise calendars. People noticed patterns in events on earth and found ways to explain such things as earthquakes and volcanoes. People have used the patterns in the strength of different materials to build structures that can withstand many natural events. Some patterns, though, are created entirely by people. What are some patterns that people create? As you continue through this unit, see what patterns you can find.

COOPERATIVE LEARNING OVERVIEW

The characters seem to be feeling pretty good about their cooperative learning skills, and they certainly have a right. Unit skills, activity skills, roles, working environments, evaluating your cooperative learning progress during wrap-up sections—these all represent a good deal of hard work.

Stop and think for a moment about how you have progressed from being new to cooperative learning to where you are now. What was the hardest part of cooperative learning? What was the best part?

- If you never practice cooperative learning again, do you have all the skills you need to be able to work cooperatively with others at any time?
- How do you think that the characters might answer the preceding question?
- Using a maximum of three speech balloons, write a conversation among the characters that you think would follow Al's comment. Write this conversation in your notebook.

Before you start the next chapter, don't forget to add your ideas to a class T-chart for the new unit 4 skill of including everyone in discussions. Remember that discussions occur in small groups (like your individual teams), as well as large groups (like your class). Also be sure to discuss with your new team of three not only the unit skill but also what you have learned in previous teams about cooperative learning.

It's Everywhere

As you look at this picture of a landfill, you might wonder what types of patterns you possibly could find. Garbage dumps, after all, contain a lot of junk jumbled together, stacked in heaps, and covered with dirt. As you might know already, garbage disposal is becoming a problem in many places. Around the world there is simply too much garbage. In this chapter you will learn about people's patterns of garbage disposal and about some patterns that people are beginning to change.

CONNECTIONS:
Look at the Evidence

Maybe when you think about garbage, it seems like something you just want to throw away. But what if you couldn't get rid of it? The garbage would keep piling up in the hallway or on the curb, and of course the mound would keep growing because you would keep using things and throwing them away.

Imagine a time when the garbage company could not pick up the garbage because of bad weather or a garbage strike. You then might have an idea of what too much garbage is like. But did you know that getting rid of garbage really *is* a problem? People in many countries are running out of places to dispose of garbage. Dumps that city planners expected to last for 100 years have filled up within 15 years. By the time you read these words, there will be even less space because of the amount of garbage that people throw away daily. In the United States, every day each person throws away an average of 4 pounds of trash. That may not seem like very much trash, but it adds up to over 185 million *tons* every year. But why should you care about garbage? It all gets thrown away *somewhere*, doesn't it?

Study Figures 14.1 through 14.3. In your notebook, record your ideas about how the garbage in each picture got to where it is.

In your notebook, record answers to the following questions.

1. What things do you usually throw away every day?
2. How much garbage (in pounds or kilograms) do you think you *really* throw away each day? Explain your answer.

Figure 14.1 ▲

Litter on the street seems ugly, but it's certainly not dangerous. Or is it?

Figure 14.2

Certain kinds of plastic products floating in the water can kill animals. Sometimes animals swim into the plastic and drown; sometimes they eat it and die. Some of the more vulnerable animals are birds and sea lions. Sea lions often swim into plastic fishing nets that fishing crews have thrown away. Roughly 50,000 sea lions become entangled and die in the nets each year. We have no estimates on how many birds die from eating plastic or swimming into it.

Figure 14.3

Sometimes when litter washes up on beaches, it causes health hazards. Sometimes things like hypodermic needles and sewage are floating in the water.

Engage ■ *Explore*

INVESTIGATION:
All in a Day's Garbage

During the 1980s and early 1990s, we could see a trend in garbage disposal. Many people have thrown away more and more garbage every year. As you read in the beginning of the chapter, some areas are running out of room for garbage. In the United States and some other countries, this is happening because our current pattern of garbage disposal is to throw away many kinds of garbage every day and in large quantities. As you and your classmates conduct the following investigation, see whether you can determine how garbage patterns have changed through the years.

Materials

For the entire class:

■ props

Procedure

1. Discuss as a class how detailed you want the production of the following play to be. You might
   - act out the play without props,
   - act out the play with props, or
   - discuss one scene per team and present the information you learned.

   If you want to use props, think about ideas and materials that you can contribute.

2. Participate in the investigation according to the procedures that you agreed on during the class discussion.

Travels with Trash

Scene 1: The 1990s

NARRATOR: Al, Marie, Isaac, and Rosalind are spending the afternoon at Isaac's house.

A nearly full garbage bag is sitting in the kitchen corner near a garbage can.

AL: Hey, Isaac, looks like you didn't take out the trash this week.

ISAAC (*offended*): Yes, I did take out the trash this week. That's just today's garbage. I can see you're not in charge of the garbage at your house, Al. You obviously have no idea how fast this stuff builds up!

ROS (*opening the tied-up garbage bag and stuffing a sack into the already full garbage bag*): Thanks for the snacks, Marie. They were really good. Hey, Isaac, looks like you guys had frozen food this week too—oops.

Working Environment

You will work as a class and then in smaller groups. As you participate, practice the new unit skill *Include everyone in discussions.*

The bag splits open, spilling plastic milk bottles, cereal boxes, frozen food boxes, and paper towel rolls.

Sorry, Isaac, I'll help clean it up.

ISAAC: That's okay; I broke a trash bag last week too.
He gets out a fresh bag.

At least you didn't see any aluminum cans, did you? I recycle those at school.

MARIE: My grandfather says that he used to take out the trash just once a week when he was growing up.

AL: Maybe he forgot to take it out on the other days.

MARIE: No, I don't think so. I think he meant there really wasn't much to take out.

ROS (*beginning to get excited*): I know how we could find out. We could go back to that time travel place and see whether we could get another free ticket for time travel. Then we could visit your granddad!

ISAAC: Well, I guess we could, but you know, Ros, the probability isn't very high of us getting free tickets twice.

MARIE: Maybe not, but I'd like to try it. I'm sure my grandfather wouldn't have made that up.

AL (*heading for the door*): Well, let's go.

Scene 2: At the Travel Agency

NARRATOR: On the street corner where Time Travel is located, the characters find a surprise.
The agency has a sign out front that states "Closed for the day. Please take a free card." Al reaches the door first.

AL: Hey, look, a free card.
Al picks one.

ISAAC: No way!

ROS (*looking over Al's shoulder and reading aloud*): We're on vacation, so why don't you take one too? This ticket is good for four stops. Just stand on the painted square, hold the card, and state the place you want to go. . . .
Ros stops and checks where she's standing.

Well, we're right in the middle of the square.

ISAAC: Now wait, I'm not sure I want to. . . .

MARIE (*moving into the painted square with the other characters*): My grandfather's house, when he was my age!

NARRATOR: And with a "poof" all four characters disappear.

Scene 3: The 1930s

NARRATOR: The characters find themselves watching a family eating breakfast.

AL: Hey, Marie, are we at your grandfather's?

MARIE: I'm not sure I recognize anybody. But if this is when my grandfather was a boy, it has to be before World War II.

ISAAC: Hey, they've got a huge wood stove. And I see chickens outside that window. We must be on a farm.

ROS: Look at all those kids!

Six children are seated around the table.

MOTHER (*standing up*): Well, you kids had better get ready to catch up on your chores today. It's time for a good spring cleaning around here. But before we start, Miguel, you take these eggshells out to the compost heap and bring me back that bowl. Ana, you take this bucket of scraps out to the pigs.

ANA: Oh, do I have to?

FATHER: Ana, obey your mother. Remember, those pigs give us food, so we should give them food.

ANA: I know, waste not want not.

Ana turns to one of the boys.

    Juan, it's your turn tomorrow!

MARIE: Hey, Juan was my grandfather's name.

Ana goes out with the bucket. Miguel goes out with the bowl.

ROS (*doubtfully*): Uh oh. Maybe that's why he took the garbage out just once a week. They all took turns.

ISAAC: Yeah, but that was a small bucket, and it was just food scraps. Besides, I don't see any plastic bags, junk mail, or anything.

Isaac and Marie walk over to Juan. Marie taps him on the shoulder.

MARIE: You know, you look a little bit like my grandfather.

ISAAC: Yeah, Juan, I know your granddaughter.

JUAN (*jumping up*): Huh? Hey, who are you kidding? Who are you?

ISAAC: Well, I do know your granddaughter—but let's skip that part. What I really want to talk about is your garbage. How come you folks don't have *more* garbage—I mean, with six kids in your family and all that? Don't you have to put out big sacks for the garbage collector to come and pick up?

JUAN: My garbage? Four strangers here to talk to me about garbage? I must be going crazy. Oh well, okay. I'll tell you about it. It seems to me like we have a lot of garbage. Every week there are food scraps for the pigs or stuff for the compost heap. And three of those kids are just visiting us today from the Mendoza farm. Nobody hauls garbage away for us. If we want to get rid of something, we haul it to the dump ourselves.

Al and Ros walk over to them and join the conversation.

AL: Really? But what about all the little stuff? You know, pudding cups, plastic bags.

JUAN: Plastic bags? Pudding cups? I don't think we have those. But sometimes we take a brown bag lunch to school. After we use a bag several times, we just burn it with the rest of the burnable trash.

MARIE: Aha! So you get rid of trash by burning it. What do you burn?

JUAN: Oh, you know, waxed paper, bread wrappers, paper sacks, sometimes newspapers after we've used them for cleaning windows or something.

AL: Yeah, I saw that stack of newspapers and it looks pretty skinny to me. (*He walks over to it and picks up a few pages.*) Where are all the ads?

ROS: So what *is* at the dump if there's no food and no paper?

JUAN: Oh, I don't know—old tin cans I guess, if we haven't used them for holding nails or something, old cars, washing machines—just big stuff, mostly. Hey, you guys wait here, and I'll take you to meet my grandfather. He knows a lot about what people used in the old days.

MOTHER (*not seeming to see the visitors*): Juan, you can start by sweeping the upstairs.

JUAN (*walking over to his mother*): Uh, could I go over to Grandpa's house first? I think he could use some of these eggs.

He gestures to a basket of eggs sitting on the kitchen table.

MOTHER: Well, all right, Juan. That's a nice idea. But I want you to come straight back here and help.

JUAN: Okay, I will.

Juan walks over to the characters.

It's okay! I can go for a few minutes.

He picks up the basket of eggs.

Scene 4: Hearing about the 1890s

JUAN'S GRANDFATHER: Hi, Juanito! What have you got there?

JUAN: Hi, Abuelo. Oh, nothing too much. I just brought you some eggs. Would you like to meet my new friends? Now what are your names?

MARIE: I'm Marie, and this is Isaac, Al, and Rosalind.

GRANDFATHER (*chuckling*): Well, Juan, you have some nice invisible friends.

JUAN: Oh, I guess you can't see them. Mom couldn't either. Well, anyway, Abuelo, I wanted to ask you about the old days. When you were little, what kind of things went out in the garbage?

GRANDFATHER: What got you to thinking about that? For one thing, I'll tell you that there certainly wasn't so much waste. I mean, newspapers were smaller and they came out just once a week, most of them. And we weren't wasting all this metal, letting cars rust away. Of course, there weren't any cars when I was a boy.

GRANDMOTHER (*coming onto the porch and joining in the conversation*): Hello, Juan! I couldn't help overhearing your conversation. And when we were children, back in the 1800s, there were no electric light bulbs, or aluminum foil, or radios—none of this modern stuff. And there wasn't a wagon you couldn't fix.

ROS: So they weren't even throwing out much metal. Hmmm.

ISAAC: Sounds like taking out the trash would've been easy back then, at least most of the time.

AL: Yeah, but maybe they had to do other things that we don't.

JUAN: Well, I've got to get back and help at home. See you later, Abuelo! Bye!

MARIE: I guess we'd better go, too. Bye, Juan. See you later.

JUAN: You will? Okay, bye.
He walks away.

MARIE: Look, here's another square to stand on.
The characters all move onto the square.

ROS: Where should we go next?

AL: Let's go look at somebody in *my* family history!

ISAAC: Okay, when?

AL: Let's just say, "Way back then!"

NARRATOR: The characters are caught up in a dark cloud, traveling far back in time.

Scene 5: The Year 1500

NARRATOR: The characters find themselves standing on a narrow street paved with stones. They are in Europe in the Middle Ages in about the year 1500.
Marie looks up. She sees that someone is about to pour something out of an upstairs window.

MARIE: Look out everybody! Move!
Just as the characters step inside the doorway, dirty dishwater comes pouring down beside them.

NARRATOR: The characters then look inside the house. They see a group of people inside, apparently a mother, father, and several children.

ROS: Hey, Al, one of those kids looks a lot like you! But what funny clothes everyone is wearing!

MARIE: I wonder what year it is?

ISAAC: Hey, I think I can hear what they're saying. It's some sort of celebration of the year 1500. Listen. . . .

WALTER: So he let you bring home a piece of paper for us?

FATHER: Yes, but he wants me to bring it back tomorrow. We mustn't try to draw on it. It's just for you to look at. It will be one of the pages used in a book for the King!

MOTHER: So this is paper. Well, I never thought I'd see the day when we'd actually have a piece of it in this house. Isabella, be careful with it! Remember, your father said it took several hours to make that sheet!

ISABELLA (*smiling*): Well, if only the King and a few other rich people can use it, I guess we won't see much of this, will we?

WALTER (*laughing*): No, I guess we'd have to have more money than we've got now!

AL: Hey, he sounds like me, doesn't he?

MARIE: Yeah, Al, he kind of does. You know, it sounds as if they didn't even have newspapers to throw away!

ISAAC: This throwing out the garbage job is looking easier all the time, but I don't like the sewage getting thrown out of the window.

ROS: No, sinks are pretty handy. So they didn't even have any paper that they would throw away. Imagine, no paper towels even!

ISAAC: Well, time to go! I guess I'll hold the card and get us home. I think I need another snack. Card, take us home!

ROS: No, Isaac—we have one more stop!

Scene 6: A Time in Prehistory

NARRATOR: Instead of ending up at home, the characters find themselves in a dim light. It seems that they might be standing near the edge of a cave. A fire is glowing in the center of the cave, and several people are eating.

ISAAC (*seeming upset*): Oh, no. Where are we now?

CAVE DWELLER 1: Hey, I wasn't done with that bird yet. Where did you throw the bone?

CAVE DWELLER 2: Well, where do you think I'd throw it? It's on the garbage heap with the other bones that have been there all year.

ROS (*noticing that she is standing near the heap of bones*): Oooh, yuck. *She moves.*

CAVE DWELLER 1 (*retrieving the bone*): Well, so much for this bone. It looks like there are ants crawling on it.
Another cave dweller is walking around with an animal hide.

CAVE DWELLER 3: Be sure you don't put that animal hide on the garbage heap. We need it for the children's clothes.

CAVE DWELLER 4: No, I wouldn't throw it away. (*He or she trips on a bone.*) You know, we might need to dig a hole for this heap or something. It's really getting in the way. I'll make a digging stick from this bone.
He or she picks up a bone from the stack.

ISAAC: That's their garbage heap for months and months? Bones?

MARIE: It looks like there are also a few arrowheads. And they even must have used quite a few of the bones, because there aren't very many in the heap.

AL: Okay, so they didn't have very much garbage. Now let's go home!

Scene 7: Back in the 1990s

NARRATOR: The characters find themselves standing in Isaac's kitchen. Beside them is another mess because the garbage bag has broken again.

ALL FOUR CHARACTERS: Oooh, yuck.

In the closing scene, the characters bend over and begin quickly picking up the "garbage" again.

Wrap Up

Write answers to the following in your notebook. Be prepared to share your own ideas if the teacher calls on you during a class discussion.

1. How was the garbage different in each of the scenes that the characters visited?

2. Describe how people disposed of garbage differently in each time period: (a) 1930s, (b) 1800s, (c) 1500s, and (d) prehistoric times.

3. List the ideas about garbage that you already know or have learned about so far in this chapter.

READING:
The Garbage Crisis

You might have heard about the garbage crisis. Maybe you have heard that we all need to recycle. You even might have participated in recycling programs in your school. But it's possible that you don't know why recycling really is necessary or why it is a good idea. If you think back to what Marie said in the connections activity Look at the Evidence, you might agree with her question.

Stop and Discuss

1. If people have been throwing things away for so long, why is there a problem now? Why do certain problems exist today that would not have existed in the past?

People have more materials to throw away today than they did 10 years ago. The more materials you have, the more you can throw away. For example, if plastic bottles didn't exist, you couldn't throw them away. It might not seem very important when you throw away a plastic bottle. Every hour of the day, people in the United States make that decision. Americans throw away over 2.5 *million* plastic bottles every *hour*.

What do we *do* with our garbage? Typically, we put it into a plastic bag and expect that the garbage collectors will take care of it. In the United States, people throw away over 185 million tons of garbage each year. That's 1,460 pounds, well over half a ton, per person. In other words, it's about 4 pounds per person, every day of the year. What might 4 pounds of your garbage consist of? If you're like many Americans, more than half of your garbage is paper and plastic. The rest of your garbage is probably food scraps, glass, metal, and lawn trash. Even if you don't have a lawn or garden, your average portion of garbage includes the lawn trash from parks.

Landfills

Garbage collectors usually haul the garbage to a **landfill** and dump it. A landfill is most often a large hole in the ground, unless it becomes so large that it is a mound. Layers of garbage are covered with dirt each day to keep the garbage from blowing away and to keep away rodents and other pests. In the United States, 73 percent of our garbage is hauled to landfills (see Figures 14.4 and 14.5).

The problem with landfills is that they are filling up. This is particularly true in some coastal areas with large populations and in the northeastern United States. People in these locations often have chosen what seemed to be the simplest solution: They have

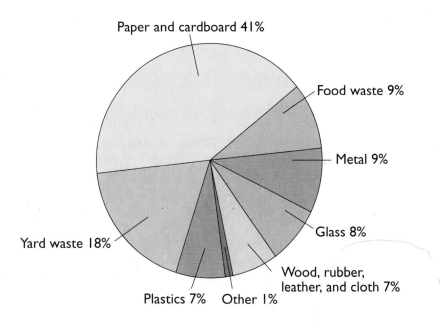

Percentage By Weight

Source: Environmental Protection Agency and Franklin Associates

Figure 14.4

This pie graph shows the composition of garbage in the United States.

Figure 14.5

This pie graph shows how we dispose of our garbage in the United States. Today, 73 percent of our garbage is hauled to landfills.

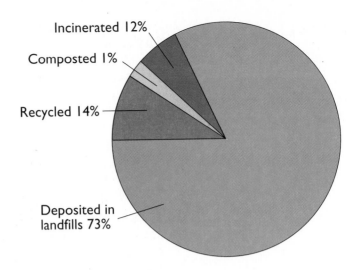

Source: Environmental Protection Agency

hired truckers or large garbage barges to haul their garbage to far-away landfills. But there are problems with using distant landfills as well. Transporting garbage for long distances is expensive, and sometimes the managers of the receiving landfill will not accept the garbage shipment. If the garbage is not accepted, the likelihood increases that the garbage haulers then will dump the garbage illegally, either in the ocean or on land—because hauling the garbage back home would be expensive. And people simply cannot find a place for the garbage back at home. For example, in New York State in 1985, there were 294 open landfills. By 1995 that number is expected to drop to only 76 landfills. And by 2006 there probably will be only 13 landfills remaining open. Clearly people in this state are running out of room for their garbage. As their garbage is hauled to other parts of the country, landfills in other places will begin (and have begun) to fill up as well. By the year 2000, half of the landfills in the United States are expected to be full.

It might seem that people just should continue to dig more and deeper landfills wherever it is possible. There are problems with this approach, however. First, many people protest if officials are considering a landfill near them. In some places where there might be room, people have put up strong protests. They don't want landfills in their backyards. This is a big change from the 1960s and early 1970s, when people didn't think of landfills as dangerous places (see Figure 14.6). Second, people cannot dig landfills much deeper than they are already. Soil is only so thick, and below it is solid rock (or underground water).

To compound the problem, leaks are occurring at landfills now. When people first began digging landfills, they were not concerned about leaks. But through the years, rain has soaked into the landfills. Because water doesn't stay in one place unless it is held by something watertight, eventually the water in old landfills moves down farther into the ground and into lakes and streams.

Figure 14.6

People once thought that garbage dumps were full of safe junk. Now people have found out that landfills are not always "sanitary."

Water that moves through the soil and underground is called **groundwater**. Groundwater that moves through landfills often becomes polluted by the material that it passes through, such as paint thinner. The danger is that this polluted water may turn out to be someone's drinking water.

Among the first people to learn about hazards from industrial landfills were the residents of the Love Canal area of New York in the 1970s. (An industrial landfill is one where manufacturing wastes are dumped.) The people living near that landfill had unusually high rates of cancer and higher than usual rates of birth defects among their children. Many people felt sick for no apparent reason. Then they discovered that they were living on top of and beside a landfill in which harmful chemicals had been dumped and then covered over in the 1950s. People recognized the strong correlation between the number of sicknesses and the old landfill. When city officials finally called in medical experts, the experts determined that some materials in the landfill were very likely causing many of the illnesses. In some homes polluted water was even bubbling up into the basements. Officials evacuated the school and many of the homes around the landfill. The residents of the Love Canal area filed a lawsuit against the company originally responsible, hoping to force it to pay for some of the damages. Even 15 years after filing the lawsuit, many of the issues of this court case remain unresolved.

Trash that is potentially harmful to people is called **toxic waste** or **hazardous waste**. Toxic means "poisonous." You would not want your drinking water to contain toxins. We define *hazardous wastes* as "things that might cause fires, corrode metals, or make people sick in different ways." If people throw away such things as motor oil, paint thinner, or car batteries, they can expect that small amounts of oil, paint thinner, or the lead from batteries might end up in somebody's drinking water. These wastes can cause sickness. When too much lead is in our drinking water, for example, lead poisoning can occur. Lead poisoning can cause mental retardation or death. People never used to worry about such things as dissolved lead or other hazardous wastes in their water. But now that we know that water can flow into and out of landfills and become polluted, people are concerned.

People have taken steps to prevent toxic chemicals from leaking into the groundwater below landfills. Modern landfills (those built since the 1980s) have plastic liners or 2-foot layers of clay at the bottom to trap any water that might fall through. Collection tanks hold the trapped water. Landfill workers then treat the trapped water to remove pollutants and sometimes circulate the water back through the landfill. When the water is recirculated, the decomposition of materials in the landfill speeds up, because moist materials decay faster. Some water probably still does get through to the ground after many years, but it is a tiny quantity compared to the leaks from old landfills.

Stop and Discuss

During this chapter you might have noticed some patterns associated with garbage disposal.

2. Using what you know from this reading so far, describe trends and correlations associated with garbage disposal. What predictions can you make about how these patterns will be different or the same in the next few years?

Incineration

Some city officials are looking for other solutions because they know that people will not allow new landfills near them and that the garbage problem will not go away. One solution is to return to burning garbage. Garbage is burned in furnaces called **incinerators**. People stopped using incinerators during the 1960s because the burning garbage produced smoke. People could see the air pollution and wanted it to stop.

In the late 1980s, new incinerators became available. These new incinerators get rid of half of the problem: They burn things at very high temperatures, and their chimneys filter out ash, so there is no smoke to see. However, they do still cause air pollution. The particles that spew from the new incinerator chimneys are simply too small for people to see, so the new incinerators do not eliminate the problem of air pollution. In addition to causing some air pollution, other problems are associated with the new incinerators. They are expensive to build and operate. It has cost some cities hundreds of thousands of dollars to hire experts who know how to run the incinerators properly. And even with the advice of experts, sometimes the incinerators break down and so leave the city without a method of garbage disposal.

When incinerators do work and city officials use them to burn all kinds of trash, another problem arises: Toxic waste becomes concentrated in the ash. Modern incinerators trap the ash that flies up the chimney and the ash that settles at the bottom of the furnace. The ash is all that remains of huge amounts of garbage. The incinerators do help reduce the amount of garbage, but some toxic waste doesn't burn. Liquids evaporate and papers are burned, but some metals are left behind. This concentration of metal can be toxic and must be disposed of in a landfill designed especially for hazardous waste.

Using incinerators does have benefits, however. First, when cities use incinerators, the landfill crisis is lessened. Second, engineers who design the new incinerators claim that they can build incinerators that release almost no poisons. In addition the heat from a certain type of incinerator can be used to generate electricity. Some countries, such as Japan and Sweden, and a few cities in the United States have waste-to-energy systems that generate electricity from burning garbage.

Stop and Discuss

3. Why do you think that some big cities have chosen incinerators as part of the solution to their garbage problem?

Recycling

Another solution to overcrowded landfills is **recycling**. Recycling means that materials repeat a cycle or, in other words, that people use them more than once. Currently people in the United States recycle only about 12 percent of their garbage. Experts estimate that we could recycle 40 to 60 percent of our garbage. But recycling requires more effort than just bagging up the trash. People must sort the trash and then possibly take it to a special location for collection. The most commonly recycled materials include aluminum, glass, paper, certain plastics, and motor oil.

Recycling has many benefits. First, it reduces the amount of garbage going to landfills and saves energy. For example, when aluminum is recycled, it produces 95 percent less air pollution, 97 percent less water pollution, and requires 95 percent less energy than mining and processing the same amount of aluminum ore from a mine. Second, because people are not simply tossing the materials away, recycling plants generate three to six times more jobs in a local area than either landfills or incineration plants. And third, many communities that require people to recycle have greatly reduced their landfill costs.

Recycling does not end with the collection of materials—it also involves purchasing. To make recycling a choice that works, people must buy recycled products. For example, if people do not insist on recycled paper, the market for it decreases. So the success of recycling depends on people both *collecting* and *buying* products made from recycled materials.

Recycling does have limits. Some materials require too much energy to recycle effectively, and others are of too poor a quality to use again. For example we can recycle paper only so many times because the wood fibers in paper have to be of a certain length. Each time the paper is chopped up to make paper pulp, the paper is weakened because the wood fibers are shortened. Also we cannot recycle paper that has glues or adhesives on it. If paper and plastic materials or different types of plastic are laminated together, they cannot be recycled either. So even though recycling is a partial answer, it does not solve the entire garbage crisis.

Reducing and Reusing

Reduction simply means that people avoid unnecessary disposable products and demand less packaging. An example of reduction would be carrying a thermos of juice to school instead of a drink packaged in a disposable drink box. Because you could *reuse* the thermos many times, you would have *reduced* the amount of garbage required to provide you with servings of juice.

Stop and Discuss

4. Some environmentalists have a slogan: reduce, reuse, recycle. Explain how this phrase could (or could not) solve the garbage crisis.

Composting

Compost is a mixture of garbage that rots and breaks down into nutrient-rich material. After this rotting occurs, garbage becomes a useful resource because it can add nutrients to soil. It might seem that all garbage in landfills would rot anyway because the heaps of garbage are just sitting there. The garbage in most landfills, however, does not rot or rots very slowly. It is buried so deeply that air cannot get to it, and even items such as leftover food and newspaper can last for over 30 years.

Materials in a compost heap decay within a few weeks or months. Composting requires the presence of several factors: air, water, warmth, and soil microbes. (Soil microbes are plants or animals that live in the soil and are so small that you cannot see them without a microscope.) To have all these factors work, compost must be kept near the surface.

Composting provides several advantages. First, composting costs very little, even on a large scale. The only costs are shredding the garbage and occasionally turning the compost so that it rots more quickly. Second, garbage that is composted becomes nutrient-rich material. This material is useful as fertilizer or as a covering for landfills. Third, if cities use this material as a covering for landfills, they will save money because they won't need to buy soil covering from rural areas or developers. Finally, and perhaps most importantly, composting reduces the amount of garbage sitting in landfills. In the United States, we could compost about 20 percent of our garbage each year.

Stop and Discuss

5. List the costs and benefits of each of the methods of garbage disposal described in this previous reading.

CONNECTIONS: Through the Years

Look at Figure 14.7. What does this graph show you? What trends, correlations, or cause-and-effect relationships can you determine from this graph? Which is increasing faster in the United States, the number of people or the amount of garbage that we throw away each year?

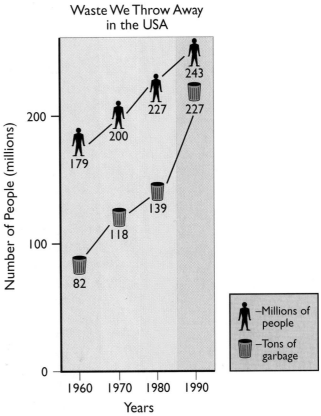

Figure 14.7

This graph shows the population in the United States from 1960 to 1990, along with the number of tons of garbage generated.

Explain

INVESTIGATION:
Pitch and Take

Garbage, garbage, and decisions, decisions. Garbage is something that you want to get rid of. Some decisions make it easier to get rid of garbage, and some make it more difficult. In this investigation you will play a game. The goal of this game is to get rid of your garbage. You and your classmates will decide on the consequences of 24 different actions before you play the game. You will travel around the board until you (or someone else on your team, depending on how you play) have gotten rid of all your garbage. As you play the game, pay attention to the actions that help you get rid of your plastic pieces. Do you suppose that those same strategies might help with the garbage crisis?

Working Environment

Work cooperatively in your team of three. Practice the skill Treat others politely. Use the role of Manager. Push your desks together or work together at a table.

Materials

For each team of three:
- 1 Pitch and Take game board
- 1 bowl for discarding plastic pieces
- 1 spinner
- 3 tokens

For each student:
- 20 plastic pieces or pinto beans (to represent trash bags)

Procedure: Part A—The Game Board

1. Obtain the materials.

2. Set aside all materials except the game board.

3. Read the actions described on the squares assigned to your team.

   Each square describes an action that someone could take. You and your classmates will decide the consequences of each action.

4. Decide whether each action will make the garbage crisis worse or better.

   Each square describes one action, such as recycling paper. Your task is to decide whether this action could help lesson the garbage crisis a little, quite a bit, or very much. If you decide that this action will make the garbage crisis worse, you are to decide whether it will make the crisis a little worse, quite a bit worse, or very much worse. Assume that 1,000 people are doing the action in each square.

5. Choose the number of plastic pieces that people are required to pick up or are able to discard if they land on that square. Decide between one, two, or three pieces.

   Each plastic piece represents a bag of garbage, and the goal of the game is to get rid of the garbage.

6. Participate in the class discussion of the game board.

Be ready to share your team's ideas about what should be in the squares.

Procedure: Part B—Playing the Game

1. Choose the token that you will use.

If your teacher did not supply individual tokens, you may use items that you have, such as a pen cap, a ring, or a coin.

2. Count out 20 plastic pieces for yourself.

All Team Members should take these from the bowl that the Manager picked up. Put the extra pieces back into the bowl.

3. Decide who will go first by using the spinner.

The high number goes first, the middle number goes second, and so on.

4. Read the rules for the game (see Figure 14.8).

5. Play the game.

Adapted from *GARBAGE* Magazine, September/October 1991 issue, Gloucester, Mass.

Wrap Up

Discuss the following questions with your teammates. Record your answers in your notebook. Be prepared to explain your answers if you are called on in a class discussion. Write your responses to the following questions in your notebook.

1. Look over the squares on the Pitch and Take game board. Do you strongly agree or disagree with any? Choose at least two squares; write down what is stated in each square; and write down why you agree, disagree, or are neutral about the information on those two squares.

2. If you could change just one thing about the garbage crisis in the United States, what would you change?

3. If you could change one thing about your use of the skill of treating others politely, what would you change?

Figure 14.8

The rules in this box describe how to play Pitch and Take.

Rules of the Game of Pitch and Take

- To take your turn, you use the spinner. The number that the spinner stops on shows how many steps your token should take.

- Move your token.

- When you land on a square, read it and do what it tells you to do. Discard or take the number of pieces that your class agreed on for that square.

- It is all right for more than one person to be on the same square at the same time.

INVESTIGATION:
Everything in Its Place

As you read in The Garbage Crisis, landfills can leak; that is, rain or snow can wash down into them, seep through the garbage, and pick up pollutants. The polluted water then may move into rivers and lakes. So water that used to be clean can become polluted, and that polluted water can move elsewhere. People who have studied relatively safe landfills have found that some soil types allow less movement of polluted water than other soil types. As you do this investigation, try to discover the answer to the following question: What kind of soil would you want underneath a landfill, and why?

Working Environment

Work cooperatively in your team of three. Use the roles of Manager, Communicator, and Tracker. In class discussions concentrate on using the unit skill. In your team of three, use the skill Treat others politely.

Materials

For each team of three students:
- 1 clear-sided plastic box with a 2.5 cm layer of soft clay or coarse sand
- 1 wood block, approximately 2.5 cm high
- 1 sprinkler container or spray bottle
- 1 soda straw
- powdered soft-drink mix without sugar (1 Tbsp or 1 packet)
- 1 bowl
- 1 graduated beaker or 1 ruler

For the entire class:
- water source
- clock

Procedure

1. Watch your teacher's demonstration of the procedure.

   In this investigation different teams will be testing soils at different sites.

2. Review with your classmates what a controlled experiment is.

3. Discuss with your classmates what each team will need to do to conduct a controlled experiment.

   Listen to the suggestions that people make for how the experiment might be set up to test your landfill sites. Your goal is to decide which soil type would be the best for a landfill.

4. Obtain the spray bottle, water, and a bowl.

5. Pump the spray bottle of water 10 times, allowing the water to spray into the bowl.

6. Measure the amount of water in the bowl.

   Use a ruler or a graduated beaker. This will allow you to be sure that all teams use the same amount of water. What other things must all teams do in the same way to control the experiment?

7. Obtain the rest of the materials.

Figure 14.9

How to set up the landfill site before adding water.

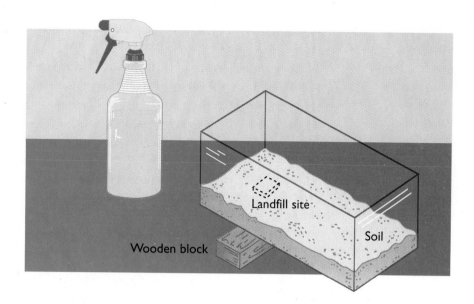

Landfill site

Wooden block

Soil

8. If the soil is not already in the box, put a 2.5-cm layer into your team's plastic box.

9. Use a straw to dig a small hole 1 cm deep.

10. Open the packet of powdered drink mix and pour all of it into the hole.

    The Communicator should do this.

11. Cover the powdered drink mix with soil.

    The Tracker should do this.

12. Tilt your box as shown in Figure 14.9.

    This is your landfill site.

13. Divide the work as follows:

    ■ the Manager holds the box and observes it;

    ■ the Tracker keeps track of time and says when to add water.

14. When the Tracker says "Go," the Communicator should add the water at the rate that the class agreed on.

    If you have a controlled experiment, all teams will be adding the same amount of water at the same rate.

15. Continue adding water and observing your landfill site for the amount of time that your class decided on.

    Notebook entry: Write down any notes that you want to remember for the class discussion, such as the results that you saw.

16. Clean up your materials as instructed.

From: Movement of Groundwater and Contaminants, which appears in Groundwater Quality Protection in Oakland County: A Sourcebook for Teachers, pp. 20–26 Copyright © 1984 by East Michigan Environmental Action Council, Bloomfield Township, Michigan. Adapted by permission.

Wrap Up

Write responses to the following questions in your science notebook.

1. In your controlled experiment, what was the one factor that varied from team to team?

2. What conditions did your class agree on so that all teams were doing a controlled experiment?

3. When you want to compare data from two or more landfill sites, why is a controlled experiment important? Give an example of when a controlled experiment could be important.

4. What type of soil did your team test? Describe what happened when you added water to your landfill.

5. What soil type do you think should be beneath a landfill, and why?

6. Describe how the unit skill helped during class discussions.

7. Decide on a reward for the teams that were the most polite.

 CONNECTIONS:
Down by the River and Out in the Desert

How might you use patterns to solve a mystery? Recall from Unit 1 that scientists often must work like detectives when they are trying to solve problems. As you proceed through this connections activity, you will look for patterns to solve two mysteries. To do this, take the following steps:

- individually, read Part A—Down by the River;

- write out your answers for the class discussion; and

- then go on to Part B—Out in the Desert.

Part A—Down by the River

Pleasant Isle is a small town along the Atlantic coast. The region is a pleasant place to live, except that it has a history of flooding. Pleasant Isle was built mostly during some unusually dry years in the 1940s, and many people didn't know about the flood patterns before they built their homes and shops. Since 1950, floods of some sort have happened almost every year. During early spring, heavy rains fall and the streams become full. Sometimes the streams overflow only slightly; sometimes the floods cover the entire floodplain. Even during the years with floods, the waters are rarely more than 30 centimeters deep. During these floods many people simply have wet yards and some water in their basements. A few people have built their homes higher on the hills, but most residents have stayed near the downtown area on the floodplain in their original homes. When new people move into the area, their neighbors usually tell them not to store anything but junk in their basements because the basements usually get flooded. Most people in Pleasant Isle think of the floods as a minor annoyance but nothing more.

Elaborate ■ *Evaluate*

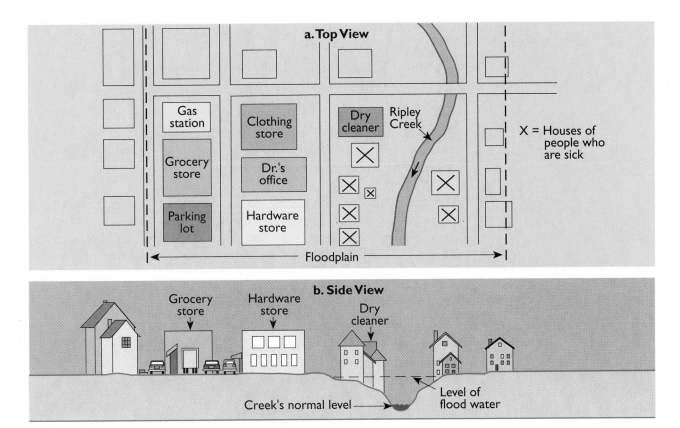

a. Top View

Gas station

Clothing store

Dry cleaner

Ripley Creek

X = Houses of people who are sick

Grocery store

Dr.'s office

Parking lot

Hardware store

Floodplain

b. Side View

Grocery store

Hardware store

Dry cleaner

Creek's normal level

Level of flood water

Figure 14.10

This diagram shows the patterns that the health officials found. The two views are a view from above (a) and a view from the side (b).

People in this region have had very few health problems. But lately, there are alarming numbers of people with kidney problems. For several years it seemed that many more people than usual had flu-like symptoms and just didn't feel well during the spring floods. Then after a few years, about 1 out of every 10 people developed kidney problems. People began talking to each other about their problems, and soon they realized that many people in Pleasant Isle were sick.

Townspeople contacted the state health department and asked for help. They learned from the officials that a number of toxic materials could cause their symptoms. Sources of toxic waste could be such things as old car batteries, paint thinner, and chemical cleaning compounds.

Health officials began to survey the area. No dumps were nearby that they could find. The community had no major industries. On the main street of the town there were several clothing stores, a grocery store, a dry cleaning store, and a hardware store. Because of hard economic times, a clothing store, the drycleaners, and the hardware store closed. During the following spring, health officials returned to the area and collected samples of the floodwater. Again people were sick. When the officials plotted the locations of sick people's homes on a map, they found the pattern shown in Figure 14.10. Then the officials began checking basements of all the buildings in town. Which do you think was the first basement they checked. Why?

1. This reading has described several patterns. What are they?
2. Once the officials completed their survey, they knew immediately what to look for. What other information did (or do) you need to solve the puzzle?

Part B—Out in the Desert

After reading Part A, you might think that a desert would be a much safer place to live than the floodplain of a river. After all, you wouldn't be near much polluted water, right? Right. But potential risks are in the desert as well.

You work for a company that advises people on how to handle hazardous waste. A member of Congress has asked your company to make some decisions about a hazardous waste dump in the desert. Currently the wastes are stored in an isolated area of Nevada. Very few people live nearby. Recent studies by geologists, however, have shown that the area has a high potential for earthquakes. The wastes are liquid and are stored in steel barrels. These are your choices:

a. Leave the wastes where they are. The probability of an earthquake affecting the barrels is 9 percent each year for each of the next 30 years. After 30 years the barrels will be slightly more rusted and more easily damaged if an earthquake occurs. Then the probability that an earthquake will affect the barrels goes up to 10 percent. If a leak occurs, there will be no way to stop it.

b. Transport the wastes to an urban waste incinerator to be burned. The probability of a traffic accident while the trucks are moving the barrels is 8 percent. In the urban area, the company could burn the wastes in a modern incinerator. When the liquid waste is boiled away, the waste would be reduced to ash. Then the company could cement the ash into concrete blocks.

   Citizens in the urban area have organized a protest against this idea. They do not want the wastes transported into or burned in their city. Furthermore they do not even want the concrete blocks of burned waste to be stored nearby. To transport the blocks away from the city would mean two-way costs for transportation. This would raise the price of this disposal method to $60 million. If the incinerator has a scrubber added to it to catch ash and reduce air pollution, the cost of the project would be $70 million. If this method is adopted and completed without accident, there would be no future fears of chemical leaks from this waste.

c. Build an incinerator near the current dump. Burn the waste that is now in the barrels and cement the ash into blocks. Because the company would need to transport many workers to this isolated area, along with the materials they

Evaluate

would need to build the incinerator, the cost for this project would be $70 million. The probability of increased air pollution is 16 percent, and the winds are strong in this area. The ash could reach a small city 200 miles away. Again the option of a scrubber is available, but this would raise the cost to $80 million. If this project is completed, there will be no future leaks to worry about.

d. Move the wastes to another isolated area with a lower risk of earthquakes. Again the probability of traffic accidents is 8 percent. The cost for this would be $2 million. This probably would prevent damage to the barrels from an earthquake, but the old barrels would rust eventually, and there would be no way to stop leaks when this occurs.

e. Place the barrels on the ocean floor because there are no earthquakes there. This would cost $6 million. The company would have to transport the barrels through a large urban area, and the probability of a traffic accident in this case is 12 percent. There still would be a potential for leaks when the barrels rusted.

Write answers to the following questions in your notebook.

1. One of the choices above contains false information. Discuss with your teammates which one you think it is.

2. Draw a costs-and-benefits table for each of the four options that do not contain false information.

3. Why do you think the hazardous wastes originally were put in the desert?

4. Which choice do you think is the best solution for the wastes stored in the desert? Justify your answer. (For this question your answer may be different from your teammates' answers.)

SIDELIGHT

How Big Is 185 Million?

How big is 185 million tons of trash? How much space does all that trash really take up? The short answer is—a lot. One year's trash in the United States takes up 18,500,000 truckloads, based on the average 10-ton load of 20 cubic yards that the average garbage truck carries. This is enough garbage to completely fill 1,000 professional football stadiums every year. What a lot of trash.

You already know that paper and cardboard products account for over 40 percent of the total weight of our trash. And, in terms of volume, paper and cardboard make up over 50 percent of our trash. So in 1 year, over 500 of the 1,000 football stadiums would be filled with nothing but paper and cardboard.

You also know that people recycle only about 12 percent of the paper they discard. That represents only 60 of the 500 football stadiums that would be filled with paper. We could be recycling 400–500 football stadium's worth of paper each year not just 60.

And that's just the paper. The other 500 football stadiums would contain plenty of trash that could be recycled—probably 50 percent of it. That would represent another 250 football stadium's worth of trash that we could recycle.

If everyone in the United States recycled as much as they could, we might have as few as 300–500 football stadium's worth of trash each year instead of 1,000.

Solving Problems

The neatly stacked tires in this photograph show one way that people are solving a particular garbage disposal problem—getting rid of tires. Used tires are a problem in many landfills. The tread on the tires has become too worn to keep using the tires on vehicles, and the tires won't stay buried because they are lighter than the dirt around them. In addition the tires can be a fire hazard because the oil that they contain is highly flammable. But some people have worked out a solution to this problem. They use worn tires as building material. In this example the tires are stacked and filled with concrete (or another weight). The tires then absorb heat and insulate these mostly underground houses against cold winters. In this case used tires have become a useful resource instead of a garbage problem.

Some garbage experts have said that we can affect the garbage crisis by the choices that we make. If you consider the tire example, this is certainly true. When people decide to use recycled products or design new ways to use recycled products, they are reducing the amount of materials that will take up space in a landfill. Although solving the garbage crisis is a complex problem, it is one that individuals can work to change. In this chapter you will discover ways that you and others can make a difference.

By now Al has been thinking about garbage for a while and so have you. For example, in Chapter 14 you read or heard about how people's patterns of garbage disposal have changed through the years. One reason that those patterns changed is because the type and amount of materials changed. The more materials people have, the more materials they throw away.

If too much garbage is a problem that people have caused, maybe people also can solve the problem. As you know by now, solving problems usually is easier when you've recognized a pattern.

Take litter, for example. Maybe you have noticed how common litter is on streets and in many parks. What are the patterns that have to do with litter? Participate in the litter hike that your teacher will organize. As you walk, record the following information in your notebook:

- all the places you find litter,
- where you find the most litter,
- what's nearby (such as a store, school, houses, or other buildings),
- the variety of litter, and
- what types of litter you see most often.

When you are done with your survey, record any trends that you find and any ideas that you have about how to solve the litter problem.

INVESTIGATION:
The Choice Is Yours—Projects, Part I

It might *seem* that your daily 4 pounds of garbage does not really amount to very much, but it does. Four pounds per day means that you throw away 1,460 pounds per year. And if you multiply your amount of garbage by 249 million (the population of the United States), then the amount of garbage is huge. But there are ways that you can make a difference. You probably know some ways already. If you don't already have ideas, this investigation will help you find some.

Materials

For each student:
- individual project materials

Procedure

1. Think of something that you would like to change about the garbage pattern in the United States.

   Notebook entry: Write this down as a goal.

2. Read through the project ideas.

   These are in the Background Information that follows this procedure. As you read, think about whether you would like to do one of the projects listed there or another one of your own.

3. Decide on a project that interests you and that is related to the garbage crisis.

   Your choice does not have to be one from the project list.

4. Decide what procedure you will need to follow to complete your project.

Working Environment

You will work individually.

Figure 15.1

This is one example of what your procedure list might look like.

Ros and Marie's Brainstorming List
Project Steps – Recycling Metal

- Write a letter to State Recycling Agency.
- Look in library for information about aluminum or steel.
- Contact local scrap metal dealers.
- Let people know where they can recycle metal.
- Find an example of recycled metal for presentation.

For example, you might decide to write letters or to look up information in the library. If you do, write down those steps. Refer to the resource list in the Background Information for other sources of information. You also might want to review How To #8, How to Conduct a Research Project, to help you remember some of the steps involved in gathering information. After you have completed your list, have your teacher check it (see Figure 15.1).

5. Prepare to discuss your proposed project with the class.

   As others discuss their projects, listen for additional steps that you will need in your project. Also be prepared to make suggestions about other projects.

6. After the discussion change the steps of your project if necessary.

7. Record in your notebook what project you have chosen and begin your project.

   Keep this in mind: You will use the information from this project to make a presentation after your project is complete.

8. Take notes as you find information and follow the guidelines in Figure 15.2.

Figure 15.2

Use these guidelines to help you decide whether you have the quantity and quality of information that you need.

Project Guidelines

- Use either four library resources or some combination of library resources and outside information (e.g., two library resources and surveys from two stores). Using a combination of sources might give you much better information for your project.

- Describe both the costs and benefits of any action that you advise.

- As you work on your project, you might decide that you need additional or different steps to complete your project successfully. Be prepared to change your project if you need to.

Background Information

Part A—Project Ideas

1. Make recycling work! Conduct a survey in several local stores to find out what recycled products they sell. Then write letters to the school board or the city council and give them information about where to buy such materials as recycled paper and plastic lumber for your school or your community. Let the city council or school board know that plastic lumber is ideal for picnic tables and outdoor playground equipment. Find and bring some examples of recycled products to your presentation. Use the library to find examples of designs that could be built with plastic lumber and what designs or uses will not work for plastic lumber.

2. Help stop the pollution caused by used motor oil! Find out who recycles motor oil in your local area. Organize a public awareness campaign to let people know how they can and why they should recycle motor oil. Let them know at least one store where they can buy recycled motor oil and one place to turn in used motor oil for recycling.

3. Which towns and cities have successful recycling programs? What makes these recycling programs work? If there are places

where recycling experts think recycling will not work well, what are their reasons? Study four locations (cities, states, or countries). Find two where a successful recycling program is operating and choose two where you think it probably would not work well at this time or where it does not yet occur but could begin quite easily. In your presentation include your ideas of what makes recycling projects successful. If no one has started a recycling program in your school, use your ideas about successful programs to start one. If a recycling program already exists, expand it.

4. What are the benefits of recycling? Study at least three types of materials. You might choose from aluminum, glass, paper, motor oil, and plastic. What are some of the effects when people do not recycle these materials? Are there materials that people should not recycle? Include examples of how people or wildlife can be affected if recycling does not occur.

5. Some communities have curbside pickup for recycling. In others people must carry their recyclables to a store or other central location. What recycling facilities are available in your community? What do they recycle? How many people use the facilities? What can you do to increase both numbers?

6. Identify 10 products that you use regularly. Do research to define and find out whether or not the products are "environmentally friendly." Find at least two or three that are available in an "environmentally friendly" form and switch to them for several days at least. In your project be sure to include how you defined an environmentally friendly product. Also describe what factors you think about when you decide whether you are willing to switch to a product.

7. Are there hazardous wastes in your home? Pick two or three specific items that you think might be hazardous, such as several kinds of batteries or different types of cleansers or solvents. If these products are toxic, find out what makes them toxic. Do a survey at your home and the homes of several classmates or an entire neighborhood. Find out how many households have these products and how people dispose of them. Look for information about safer products that people could use. Write a commercial that tells people about safer products that they can use instead of toxic ones or how they can dispose of some hazardous materials safely.

8. What are some effects that landfills have on animal and plant populations? If possible, study the wildlife in an area such as a canyon used for a landfill. Also make use of library resources to learn more about wildlife in landfill areas. If you discover that a landfill is having a negative effect on local wildlife, develop an advertising campaign. Convince people to recycle or to

Explore ■ *Explain*

produce less garbage in order to preserve the wildlife habitat. If you discover that the landfill is beneficial for some animals or plants, you might decide to describe that effect too.

9. Start a compost project at home or school. What arrangements would you need to make to get your project started? What patterns do you need to change? Are people throwing away food scraps that they could put into compost? Once the compost is ready, how would you use it? Would it be practical for your school to compost food wastes? Build a container for small-scale composting. Bring it to class and describe how you would use it.

10. Some scientists state that we are putting garbage into the atmosphere. One example is the propellants that we put into the atmosphere when we use aerosol sprays. How can you find out what aerosol propellants people commonly use? What are some of the effects of using aerosol sprays? What are some alternatives to using aerosols?

11. How might incineration play a part in solving our garbage crisis; that is, what are the methods of incineration, which methods are best, and why should we use them? Why do some people want incinerators to be shut down? Prepare a poster that describes how modern incinerators work. Choose one type of incinerator that you would consider to be the best designed; that is, describe the incinerator that offers the best technology in your opinion. After your research describe what part incineration should or should not play in garbage disposal. If appropriate, consider sending a summary of your findings to the city council.

Part B—Resource List

The following organizations and resources might have information that you are looking for. Your teacher also has a list that includes names and addresses for some national sources, and you can use the phone book to contact local organizations yourself.

The Recycler's Handbook. This book lists recycling ideas and includes names and addresses of state recycling agencies, as well as many national organizations involved in recycling.

Iowa Department of Natural Resources. This organization provides information about household hazardous wastes.

Partnership for Plastics Progress. This organization publishes information about recycled plastics as waste and about plastics recycling.

Bureau of Mines. This branch of the federal government is interested in recycling metals and in retrieving lead from batteries, among other projects.

Recycled products catalogs. Your teacher has addresses for sources of such materials as carpet made from plastic, rechargeable batteries, nontoxic cleaners, plastic lumber, and recycled paper.

Local landfill operators or scrap metal dealers. Consult these people for information about the efforts of your local merchants to promote recycling, as well as the difficulties they have encountered.

Wrap Up

Write a one-page summary of your project in your notebook. Be as thorough as you can. Describe what you did and what you learned from your project. Also describe any decisions that you made about actions to take and the costs and benefits of these actions. (An example of a personal action would be to buy an environmentally friendly product rather than one that is not. Your action should match the type of project you did.) At the end of your summary, list the sources you used.

INVESTIGATION:
Presentations—Projects, Part II

The way you present an idea can have a big impact on how people receive it. Before you can convince people of the value of your idea, you must understand clearly what you want to say and what you want to show to your audience. For example how many times have commercials convinced you that you wanted to do something or at least made you start thinking about it? In this investigation you will find ways to present your ideas and to let people know that solutions to the garbage crisis really matter. The purpose of this investigation is to prepare and present what you discovered during your project.

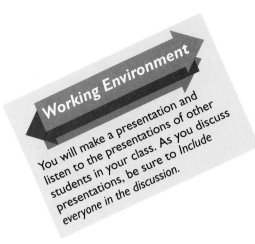

Materials

For each student:

■ materials for presentations

Procedure

1. Think about your project and what ideas you would most want people to remember.

2. Decide what information you will present to your class.

   Look over the summary that you wrote about your project and review your findings as well as how you gathered information.

3. Prepare your project presentation.

   Think of your audience and remember that they don't know all the details of your project the way you do. No matter what format you use, your presentation must include the following information:

   ■ *a description of what you chose to study,*

   ■ *how you gathered your information (e.g., whom you surveyed and what library books or magazines you used), and*

   ■ *the argument that you are making (what you want to let people know).*

4. Practice your presentation.

   Make it as appealing as possible.

5. Make your presentation and listen to your classmates' presentations.

6. As you listen to other presentations, write a summary of each one. Use two or three sentences to answer the following:

   ■ What project did the student(s) choose?

   ■ What information did the student(s) gather?

   ■ What were the results?

Wrap Up

After you and all of your classmates have given presentations, answer the following in your notebook.

1. Describe in a few sentences which project you liked the best and why.

2. How would you change or improve your project if you did it again?

3. Write down which project ideas you think you could carry out and which ones you think you couldn't. Explain what would make some ideas possible and some more difficult.

CONNECTIONS:
St. Louis and San Diego: Opposite Decisions?

As you might have noticed during this unit, if people are concerned about garbage, they must make careful choices about the products that they buy and use. Many people want to use products that cause the least pollution or that have the least negative impact on the environment. Many grocery store owners are faced with a dilemma of what to supply to their customers: paper or plastic bags.

Sometimes two people might be trying to solve the same problem and yet they might make two different decisions. If they make different decisions, can *both* be helpful to the environment?

Read the following excerpts and see what you think. Two grocery store owners are telling about their decisions. Schnucks is a chain of grocery stores in St. Louis, and Big Bear Markets is a chain in San Diego.

Why Recyclable Plastic Bags Make Sense
by Terry Schnuck

Two years ago, we introduced polyethylene plastic grocery sacks in our stores. We did so for three basic reasons. First, the bags we chose are photodegradable (they break down in the presence of air and sunlight). Second, the manufacturer uses water-based inks. Using this type of ink means the bags cannot cause harmful leaching when placed in landfills. Third, we felt the bags represented a quality product at a reasonable cost.

Since introducing the plastic bags we have found they offer many other benefits. Schnucks' environmental philosophy, which we communicated to the public on Earth Day 1990, is comprised of three critical points—reduce, reuse, and recycle. Plastic bags can address all three:

- They require less energy to produce than paper bags.

- They are reusable in the home.

- They are completely recyclable.

In April, we placed recycling containers at all of our 59 supermarkets. When the containers are full, the bags are transported from our stores to a central warehouse where they are baled for shipment to our bag supplier, Sonoco Products.

The bags are then used to make a variety of new products such as irrigation pipes, protective edge molding, plastic trash cans, sign posts and non-food packaging materials.

Even if the bags are not recycled, but instead end up in landfills, they take up only one-seventh the space of paper bags. They also burn cleanly when incinerated.

Some scientists have come to regard plastic bags as less damaging to the environment than paper bags. Papermaking contributes to water pollution, acid rain, releases dioxin and depletes our forests.

Our customers have come to prefer plastic bags (three to one over paper) for a variety of reasons. Elderly customers and customers who have to walk up stairs prefer plastic because the handles make them easier to carry. Other customers have told us the bags are great for all types of reuses because they are waterproof. But their single most redeeming feature is that they can be recycled.

Customer response to this recycling effort has been extremely positive. People want to help, and welcome an opportunity to address the solid waste problem in even a small way.

Yet the very nature of our business makes us incompatible with recycling on any major scale. Bringing products for recycling to a food store presents a potential sanitation problem.

For this reason, we have placed the bins for milk and soda containers a considerable distance from our stores on our parking lots; there is somewhat less of a hazard with the plastic bags. Nevertheless, we are pleased to be part of both these recycling programs.

There are programs companies can institute in-house as well. At Schnucks, we have reduced the number of polystyrene foam cups our office personnel use by at least 2,000 per week. We asked them to bring their own coffee mugs and drinking cups in exchange for a lower price for a cup of coffee. We also have a receptacle in our cafeteria for aluminum cans, which volunteers then take to recycling centers.

Finally, as a visible and involved member of the communities in which we do business, Schnucks strives to present a balanced view on environmental issues. For example, when talking with customers, we make it clear that our plastic bags are not biodegradable, but only photodegradable.

That is to say, we try to explain that the bags will degrade if exposed to sunlight and air—which addresses the litter problem—but they will not degrade in a landfill. However, it is equally important to point out that the bags are non-toxic and take up significantly less space in landfills than paper bags, and that they can be recycled.

From "Why Recyclable Plastic Bags Make Sense," by Terry Schnuck, *The Progressive Grocer Special Report: The Environment*, © 1990 by The Progressive Grocer. Excerpted with permission.

Why We Switched Back to Paper Bags
by Tom Dahlen

At Big Bear Markets, we feel it is our responsibility to be the environmental conscience of our customers. We must provide them with the information to make environmentally sound shopping decisions that can have a positive impact on our community and quality of life.

The issue with Big Bear is not so much paper vs. plastic, but the specific challenges that must be faced to protect San Diego's environment. There is a serious landfill shortage and the health of wildlife and marine life to consider.

While both paper and plastic bags have some trade-offs, experts from the Audubon Society have pointed out that paper bags are preferable in coastal areas because plastic bags can entangle and trap marine life. Paper bags—which quickly degrade if they find their way to the sea—do not pose this danger. Some would argue that plastic bags can degrade as well, but as we have all seen the issue of degradable plastics has become extremely controversial.

By educating our employees and our customers about the paper bags, we have attained greater productivity at the checkstand and savings in supply costs. Our customer research indicated that we had been wasting 10% of each bag—which were rarely filled to capacity. So besides having our bags made out of recycled paper, we had our supplier make them 3 inches shorter than traditional sacks.

This has resulted in better loading, easier carrying and less likelihood of toppling over in the car or when being unloaded at home. We also are paying consumers 2 cents for every large paper bag returned to our stores to pack subsequent orders.

We let the customers vote, and don't make them feel guilty about their choice. While we don't mind working with plastic, we do ask customers, "Is paper okay today?" About 95% of the time the answer is "yes."

We are also selling recycled paper bags with handles, and donating proceeds to I Love a Clean San Diego. The bag supplier has committed to planting three seedlings for every tree they cut down.

The initial switch to paper did have its costs. For example, we spent $7,000 retraining our employees to pack paper bags properly. And we spent another $20,000 in advertising to educate our customers about the change—and the reasons behind it. There was also some expense involved in removing the plastic bag racks and printing promotional material.

But six months after the switch, we have toted up some significant benefits. Supply costs are down $35,000. Big Bear is using fewer bags per unit of retail sale since paper bags hold more product. This accomplishes our goal of using up fewer natural resources.

In addition, productivity studies indicate a significant improvement in checkstand speed—over 15% on average. Finally, paper bags are easier to use and in the long run require far less bagger training. Since we are faced with increased labor costs, this is a significant cost advantage.

Our paper bags represent recycling in action. Our old corrugated boxes return as carryout bags. All in all the response to our switch to paper bags has been overwhelmingly positive—the amount of media coverage and support phenomenal.

From Tom Dahlen, "Why We Switched Back to Paper Bags," *The Progressive Grocer Special Report: The Environment,* © 1990 by The Progressive Grocer. Excerpted with permission.

Wrap Up

Read through the following and write your answers in your notebook. Prepare to share your ideas with the class.

1. Describe how Big Bear Markets changed a pattern in its customers' use of grocery sacks.

2. Describe the customers' patterns of using grocery sacks at Schnucks' Markets.

3. What do the owners of Schnucks' Markets consider the costs and benefits of their decision to be?

4. What do the owners of Big Bear Markets consider the costs and benefits of their decision to be?

5. Some people think that paper or plastic is not a choice they need to consider because they use cloth bags. What do you think the costs and benefits of this decision would be?

6. Write a paragraph expressing your opinion of the decision the Schnucks Markets made, the decision the Big Bear Markets made, and the paper versus plastic controversy.

7. What would you decide to do if you owned a large chain of grocery stores?

S I D E L I G H T

Closing the Loop

You might have seen the recycling arrows on many products. These recycling arrows stand for the three phases of recycling: collecting, remanufacturing, and remarketing.

When people collect, remanufacture, and resell products such as aluminum cans, manufacturers say that *closed-loop* recycling has occurred; that is, the product made it all the way around the loop. People sometimes forget the third part of the loop: *buying* recycled products. People who recycle paper have a saying: "If you're not *buying* recycled products, you're not recycling!"

Why do you suppose recyclers say this? It is because of supply and demand. If there is no *demand* for products that have been recycled, the *supply* will dry up. Recyclers will go out of business because they cannot sell their products. If no companies recycle products, then there will be nowhere to take used aluminum, glass, cans, or paper. These products will end up in the garbage along with the other kinds of wastes.

When you measure the length of someone's foot with a ruler, the measurement you get will not always be a whole number. In fact the measurement is more likely to be *between* two whole numbers on the ruler. So which whole number should you write down as the measurement of the person's foot?

How did Isaac, Ros, and Al decide to round off the measurements they made? They made use of **millimeters (mm)**, the small spaces between the **centimeter (cm)** marks, on their rulers. (The centimeter marks are the whole numbers.) See Figure H1.1.

Each centimeter contains 10 millimeters. In other words each millimeter equals one-tenth of a centimeter. That relationship (10 millimeters = 1 centimeter) makes it easy to show measurements

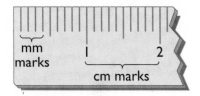

Figure H1.1

How many millimeters are between the centimeter marks on a metric ruler?

that fall between two centimeters. All you have to do is write the number of millimeters as a decimal following the whole number of centimeters. After you work through some examples, you will understand how to show measurements using centimeters and millimeters.

Let's say you measured the length of a toddler's foot and the measurement was *between* 11 and 12 centimeters—the exact measurement was 11 centimeters plus 3 millimeters. Because each millimeter equals one-tenth of a centimeter, you can represent 11 centimeters and 3 millimeters as 11.3 centimeters. (This is the same as saying the toddler's foot is 11 and three-tenths centimeters long.) Another toddler's foot might be 11 centimeters plus 8 millimeters in length, which is the same as 11.8 centimeters, or 11 and eight-tenths centimeters long (see Figure H1.2).

Now you know how to write measurements to the nearest millimeter. Your assignment in the investigation Getting Off on the Right Foot, however, was to determine the length of your foot to the nearest *centimeter*. To do this, you must **round off** your measurement to the nearest whole number. For example, if your foot were 20.2 centimeters long, that measurement is closer to 20 centimeters than to 21 centimeters, so you would round off your measurement to 20 centimeters. If your foot measured 20.7 centimeters, that measurement is closer to 21 centimeters than to 20 centimeters, so you would round off your measurement to 21 centimeters. Look back at the picture of Isaac, Ros, and Al and review their decisions about rounding off their measurements.

Figure H1.2

What are the exact measurements of two toddlers' feet that are between 11 and 12 centimeters long?

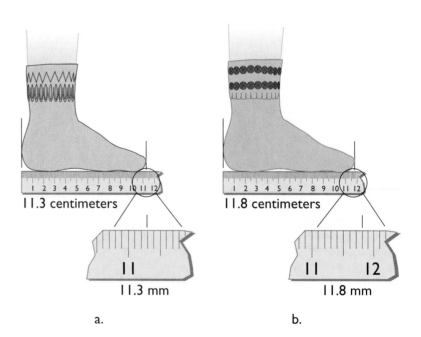

11.3 centimeters

11

11.3 mm

a.

11.8 centimeters

11   12

11.8 mm

b.

Oh, I get it! Three millimeters is the same as 0.3 cm and 8 millimeters equals 0.8 cm. Now, if I could only pin my little sister down long enough to measure her foot.

In summary, to round off a measurement to the nearest whole number, first look at the ruler. If the measurement is closer to 20, write 20. If it is closer to 21, write 21. If the measurement is exactly halfway, for example 20.5 centimeters, then round the measurement to the next higher number, in this case to 21 centimeters.

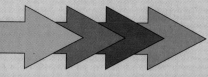
Graphs are useful tools because they help us see a lot of information at one time. We organize graphs in a certain way so that we can "read" easily the information the graphs contain. Graphs also help us find patterns in the data we have collected.

A graph has two lines, one that runs crosswise (horizontally) and one that runs up and down (vertically). These lines have special names. We call the line that runs crosswise the **horizontal axis** and the line that runs up and down the **vertical axis**. The point where these two lines, or **axes,** meet is the place where the graph begins. (The term **axes** is the plural for the term **axis.**)

Figure H2.1

The names of the axes on a graph are the horizontal axis and the vertical axis.

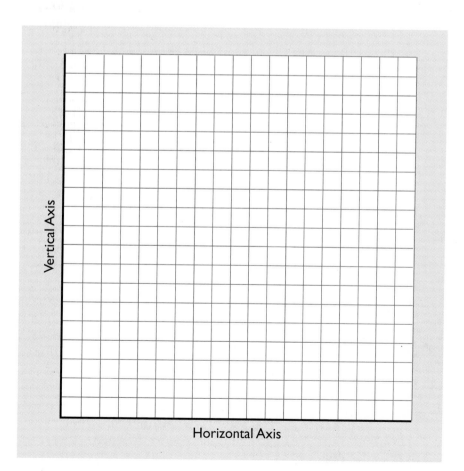

Figure H2.2

These are possible labels for the sample graph in Getting Off on the Right Foot.

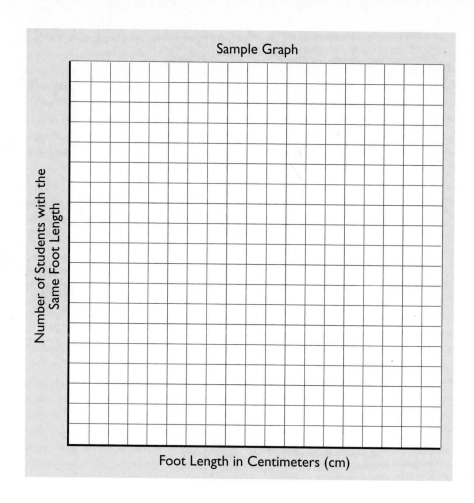

Before you began the investigation Getting Off on the Right Foot, your teacher probably set up a horizontal axis and a vertical axis for your class graph.

To find out what kind of information a graph displays, you need to look at the labels next to each axis. Your teacher probably used the following labels for the graph in the investigation Getting Off on the Right Foot.

Both axes of a graph often have a sequence of numbers called a **number scale**. You read the numbers on the horizontal axis from left to right and those on the vertical axis from bottom to top. For example, in your class graph for Getting Off on the Right Foot, the number scales probably looked like those in Figure H2.3.

The numbers on the horizontal axis represent possible foot measurements; the numbers on the vertical axis show how many students had the same foot measurement. The difference between numbers next to each other on the horizontal axis is one; the difference between numbers next to each other on the vertical axis is also one.

The number scale on one axis does not have to be *exactly* the same as the number scale on the other axis (see Figure H2.3). The difference between the numbers next to each other on an axis must be the same, though. Look at the graphs in Figure H2.4. One graph has a number scale that will work, and the other has a number scale that will not work. Which graph's number scale will not work? Be prepared to explain why you chose the one you did.

Often you will draw your graphs on a special kind of paper called **graph paper**. Graph paper is special because it has evenly spaced lines: The distance between the lines is the same from side to side and from bottom to top. This even spacing helps you show that the difference between numbers on your graph's number scale is the same from one number to the next (see Figure H2.5).

One last important feature of a graph is its title. We gave the graphs in this How To the titles of "Sample Graph" and "Record of

Figure H2.3

In addition to the labels, this diagram shows the number scales for the sample graph in Getting Off on the Right Foot.

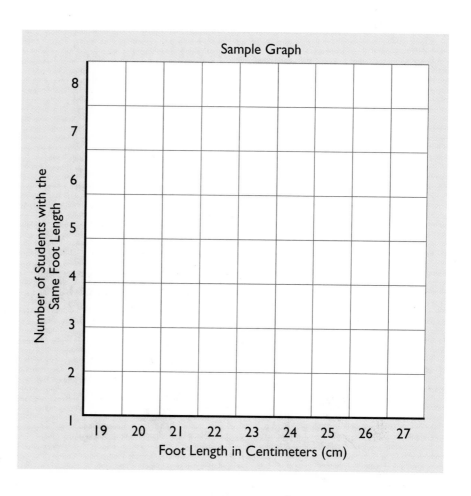

Figure H2.4

Which graph's number scales will not work for graphing data?

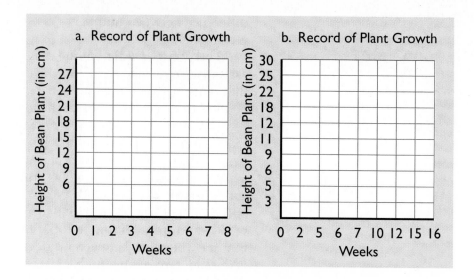

a. Record of Plant Growth   b. Record of Plant Growth

Figure H2.5

Can you explain why the example shown in B is not graph paper?

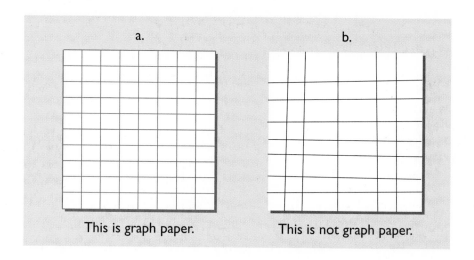

a.   b.

This is graph paper.   This is not graph paper.

Plant Growth." Your class graph for the investigation Getting Off on the Right Foot might have been titled "Class Graph of Foot Lengths." The title should tell the reader something about the purpose of the graph.

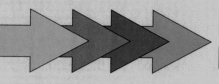
If you have read How To #2, you know that graphs have the following parts: a horizontal axis, a vertical axis, labels on each axis, a number scale on each axis, and a title. (If you have not read How To #2 or if you do not remember the parts of a graph, review How To #2 before you continue with this How To.) You also know how to organize those parts so that the graph makes sense. But no graph can really make sense until you plot some information, or data, on it.

How do you get the information, or data, onto the graph in the right places? The answer to that question is what this How To is all about: learning how to plot data on a graph. On graph paper draw a graph and plot the data from each of the sample data tables. The following steps will help you, and some hints and checkpoints appear along the way.

Figure H3.1

This data table shows daily high temperatures in degrees Fahrenheit (°F) over a 2-week period in Phoenix, Arizona.

Daily High Temperatures in degrees Fahrenheit °F, Phoenix, Arizona	
Day, Week 1	**Temperature (in °F)**
1	103
2	104
3	105
4	104
5	105
6	102
7	102
Week 2	**Temperature (in °F)**
8	103
9	105
10	101
11	102
12	104
13	100
14	99

Part A—The Steps

1. Review the data you have recorded.

   For this example, you will use the data we recorded in Figure H3.1.

2. Draw the horizontal axis and the vertical axis for the graph.

   It helps if you draw your graph on graph paper because graph paper provides evenly spaced lines and squares. The spaces on graph paper keep the distance exactly the same between the consecutive numbers on each axis, that is, the numbers next to each other on an axis.

3. Label each axis by using the headings in the data table (see Figure H3.2).

4. Set up the number scales on each axis.

   Be sure to provide space on the horizontal axis and on the vertical axis for all the numbers that are included in the data table. Remember, a number scale does not always have to start with the number 1 (see Figure H3.3).

5. Give your graph a title.

   You might title this graph "Daily High Temperatures (in °F) over a 2-week period in Phoenix, Arizona."

Figure H3.2

These are possible labels for a graph of daily high temperatures.

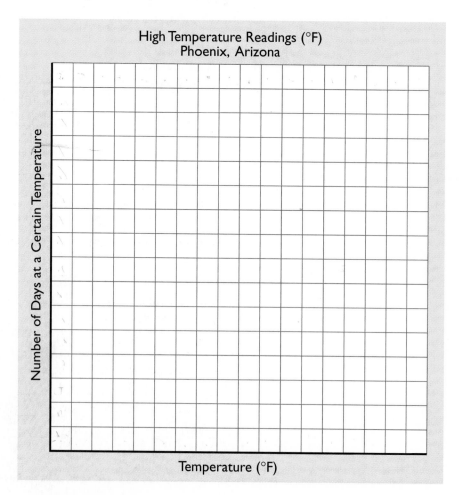

High Temperature Readings (°F)
Phoenix, Arizona

Number of Days at a Certain Temperature

Temperature (°F)

Figure H3.3

This outline shows the labels
and number scales for a graph
of daily high temperatures.

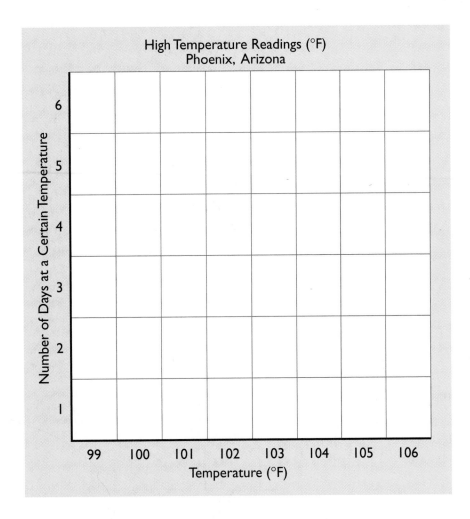

High Temperature Readings (°F)
Phoenix, Arizona

6. Plot the data on your graph by doing the following:

■ read one piece of data from the data table,

■ find the square on your graph where that piece of data fits, and

■ place an *X* in that square.

Repeat these steps for all the pieces of data in the data table.

First locate the first piece of data in the data table. It tells you that the high temperature on Day 1 in Phoenix, Arizona was 103°F. Next find 103 on the horizontal axis of your graph. Then place an X in the first square above 103.

This data point indicates to anyone reading the graph that the high temperature on *one* day during the 2-week period in Phoenix, Arizona, was 103°F. (Note that on this type of graph you will not be able to show that the temperature on Day 1 was 103°F. What you will show when you complete the graph is *the number* of days that had a high temperature of 103°F.

Go to the second day. The data indicate that the high temperature on Day 2 was 104°F. Find the first square above 104°F on the horizontal axis and place an X in it.

Plot the data for the first 4 days. Your graph should look now like the graph in Figure H3.4. Note that when you have two data points that are the same—104°F, for example—that the graph shows two Xs in that column.

Part B—Additional Practice

Suppose that someone recorded the daily high temperatures in degrees Fahrenheit (°F) in Phoenix, Arizona, for a third week. Figure H3.5 shows the high temperatures during the third week.

Plot the additional data on your graph and then answer these questions.

1. What was the most common temperature over the 3-week period?

Figure H3.4

Does your graph look like this one?

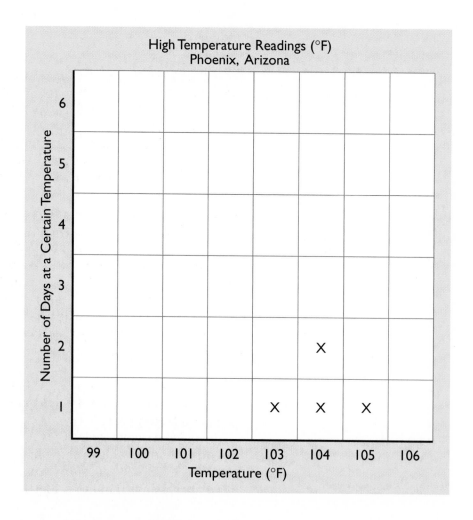

Figure H3.5

This data table shows daily high temperatures for a third week in Phoenix, Arizona.

| Daily High Temperatures in degrees Fahrenheit (°F) ||
Day, Week 3	Temperature (in °F)
1	100
2	99
3	103
4	101
5	101
6	99
7	102

2. On how many days was the high temperature 104°F?

3. During the 3-week period, did the temperature reach a high of 102°F more often than it reached a high of 99°F?

Part C—More Practice with Bar Graphs

So far in this How To you have learned how to plot data on a graph by placing Xs in the appropriate squares on the graph. In doing this you have made a type of graph called a **bar graph**. You might see why this type of graph is called a bar graph if you shade each square that contains an X. The shaded squares line up on top of one another to make bars.

Sometimes in creating bar graphs, though, you will not always have data that you can plot by marking one X at a time in a square on your graph. Sometimes you will make a bar graph by shading the entire bar at one time. If you work through the following example, you will find out how to make a bar graph by shading each bar all at once.

Look at the data table in Figure H3.6. In this data table, you will find the names and approximate heights of some mountain peaks from around the world. (The column label reads "*Approximate Height*" because the numbers are not exact. We rounded off the numbers to make them easier for you to plot.) By plotting these data on a bar graph, you can compare the heights of these mountains. Can you use an atlas to find the location of each mountain?

Remember the six steps you used in the first example? You will follow the same steps in this example, but step 6 introduces a few changes. The steps are:

1. Review the data.

   Use the data recorded in Figure H3.6.

2. Draw the axes.

3. Label each axis.

   Because vertical bars usually are easier to read than horizontal ones, you should label the vertical axis "Approximate Height (in m)" and the horizontal axis "Mountains." This way, the bars will run up and down and not across the graph. Can you see what would happen to the bars if you exchanged the labels on the axes?

4. Set up the number scale on each axis.

   In this graph, only the vertical axis has a number scale. The horizontal axis uses names instead of numbers. With some graphs, names instead of numbers will be more appropriate as labels. Also place the numbers on the vertical axis next to the lines, not between the lines (see Figure H3.7).

Figure H3.6

Which bar will be the tallest on your graph?

Approximate Heights (in meters [m]) of Mountains around the World	
Name of Mountain	**Approximate Height (m)**
Kilimanjaro	5,900
Mount Cook	3,800
Mount Elbrus	5,600
Mount Everest	8,800
Mount Fuji	3,800
Mount McKinley	6,200
Mount Whitney	4,400
Orizaba	5,700

Figure H3.7

Find the location on your bar graph for the first piece of data.

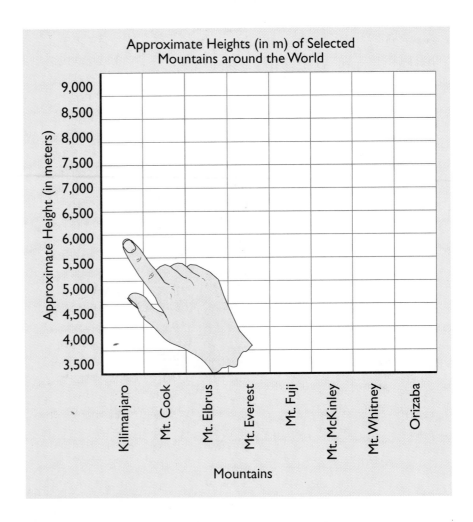

Approximate Heights (in m) of Selected Mountains around the World

5. Give your graph a title.

You might title this graph "Approximate Heights of Mountains around the World."

6. Plot the data.

The parts of step 6 that you followed before do not fit the graph for approximate heights of mountains because you cannot plot each piece of data in the data table as one X on your graph. To make this graph, you must draw the entire bar to show the approximate height of each mountain. The following steps will tell you what to do differently.

■ Read one piece of data from the data table.

The first piece of data tells you that a mountain named Kilimanjaro is approximately 5,900 meters high.

■ Find the label for that piece of data on the horizontal axis.

The label on the horizontal axis for this piece of data is "Kilimanjaro."

■ Trace your finger up the column above the label to the place on the vertical axis that shows where that piece of data fits.

Figure H3.8

This bar graph shows the first colored bar. You can do the rest.

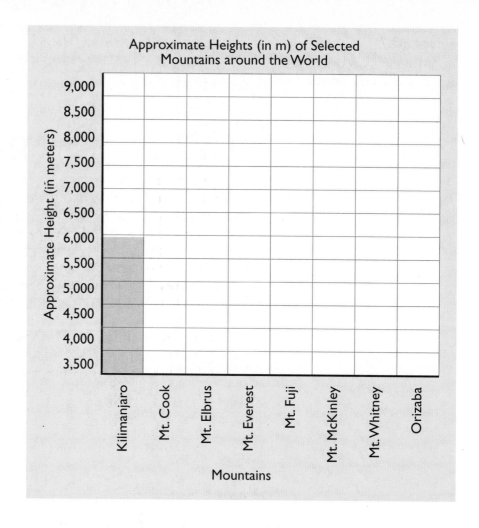

The height of Kilimanjaro is approximately 5,900 meters. That height is between 5,500 meters and 6,000 meters on the vertical number scale. You can see in Figure H3.7 that the top of the bar for Kilimanjaro is between those two marks, but much closer to 6,000 meters than to 5,500 meters.

- Draw a horizontal line at the correct height to make the top of the bar.
- Color in the bar from that line down to the horizontal axis (see Figure H3.8).
- Repeat the parts of step 6 for all the pieces of data in the data table.

Now you know two different ways to plot data so that you can make a bar graph. In How To #6, you will learn how to plot data to make a line graph, another type of graph that displays data by using lines instead of bars.

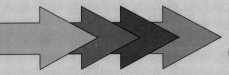

Whenever you work on an activity in this program, you collect information. For example in the first unit you collected information about things such as the length of classmates' feet; the lengths of your ears, forearms, and little fingers; plant growth; and floating and sinking objects. We refer to such information as **data.** Data can be things such as measurements, numbers, observations, dates, or times. (We say "data are," not "data is," because the word *data* is plural. The singular form is *datum,* a word that means "one piece of information.")

Usually data do not make sense until they are organized in some way. In chapters 1 through 5, the directions in your book helped you organize your data. But beginning in Chapter 6, you

won't get as much help from your book. For example, when you worked with maze puzzles in the first investigation in Chapter 6, your book did not tell you how to keep track of your observations about the maze puzzles. The instructions suggested that you and your partner test your ideas and record your results, but the instructions did not tell you *how* to record what you observed. You might have run several tests before you and your partner realized that you needed to keep track of the observations as you made them. How did you keep track of the data from your tests? Were your data organized like Al's?

Al definitely needs a better system for organizing his data. What might you suggest to Al so that he is better organized next time? You might suggest that he organize his data in a **data table.**

A data table is a chart that helps people keep track of observations, or data, so that they can use the data later to make decisions, predictions, or graphs. You probably are familiar with data tables because you used them throughout Unit 1, but you might not have tried to construct a data table on your own. It's not too hard if you have some steps to follow.

Let's suppose that one of your tasks in the investigation Maze Puzzles was to construct a data table. How might you have done it? The following steps will show you one possible way. The steps also can help you construct data tables for other investigations, such as The If–Then Box.

Why do we need to read about constructing a data table for Maze Puzzles? We already finished that investigation.

If we use a familiar example, maybe the steps will make more sense. We can find out what we could have done to be more organized. Then we can be more organized next time.

1. Read the entire investigation so that you know its purpose.

   As you read the investigation, ask yourself, "What am I supposed to find out in this investigation? What question am I trying to answer?"

   In the investigation Maze Puzzles, you were supposed to decide what the inside of your and another team's maze puzzles looked like, without actually looking inside the puzzles.

2. Make a list of the types of information that you have or that you will need to collect during the investigation.

   In the case of Maze Puzzles, especially if you and your partner did not agree about what was inside the maze puzzle, you needed information about the following:

   - which test you were doing: Test #1, Test #2, and so on
   - what you did in each test: Which way did you move the maze puzzle—up and down, from side to side, from left to right?
   - what happened when you conducted each test: Did the marble stop in a particular place? Did the marble roll freely along the edge?
   - your conclusion about what happened: Do your results mean that a straight wall is probably inside the puzzle? Do your results show that the marble probably goes in a circle?

3. Review your list from step 2 and the purpose from step 1 and decide whether you are collecting enough information.

   This probably is the hardest step. You might ask yourself, "If I collect only what is on my list, will I collect enough information to find the answer to the main question?" Remember that it is better to have too much information in a data table than too little. If you don't record all the information you will need, then you might have to repeat parts or all of the investigation to get the information you left out or forgot. This step will become easier as you construct more data tables.

4. Construct the outline of your data table by drawing one column for each item on your list.

   For the investigation Maze Puzzles, you would have had four things on your list: (1) which test, (2) what you did, (3) what happened, and (4) your conclusions. Therefore you would have needed four columns in your data table.

5. Label the columns in your data table.

   For Maze Puzzles you would have used the items in your list from step 2 as the column labels (see Figure H4.1).

6. Give your data table a title.

   The title should be simple but one that makes this data table different from any other data table you might use. A title also gives you a handy way to refer to this particular data table. You might have used

"Maze Puzzles Data Table" as the title of your data table for the investigation Maze Puzzles.

7. Fill in your data table as you do the investigation.

Figure H4.2 shows an example of what your data table might have looked like after you conducted two tests of one maze puzzle. (Your results might or might not agree with this example. The sample data table does not contain actual data, only examples of what the data might have been.)

Be sure to complete one row of the data table after each test. If you write your observations and results right after you complete each test, then you do not have to worry about remembering what you did and what happened. But if you wait to record your observations until you have completed three tests, you might not remember exactly what you did in each test and what happened.

If you had made a data table in the investigation Maze Puzzles, would you have followed steps similar to these? If not, what would you have done differently? Would your method have worked just as well? Take a few minutes to share your strategies with your classmates. You might find that the steps you would have followed are slightly different from these. That is all right, as long as you would have ended up with a data table in which your data were organized and easy for you and others to read. The true test is

Figure H4.1

This is a possible outline of a data table for the investigation Maze Puzzles.

Test number	What we did	What happened	Conclusion

Test number	What we did	What happened	Conclusion
1	We held the puzzle up so that the marble was at the bottom. Then we slowly turned the puzzle in a clockwise direction.	The marble rolled along the bottom about 1 inch and then stopped.	There is a wall or part of a wall on the right side of the maze.
2	We tipped the puzzle so that the marble was resting at the top of the maze puzzle. This time we turned the puzzle counterclockwise.	The marble rolled about an inch again and stopped.	There is a wall that stops the marble on the right side.
3	We tipped the puzzle so that the marble was at the bottom and as far as it would go to the right. Then we tipped the puzzle down.	The marble rolled straight to the top.	Now we're <u>sure</u> there's a wall on the right side.

Figure H4.2

This sample data table shows possible results after testing one maze puzzle.

whether you can use the information in your data table easily to help you answer the main question of the investigation. If your data table does not help you, then either you did not organize your data well or you forgot to collect and record some important data.

Don't worry if the first data tables you construct aren't perfect. You will have other chances to practice constructing data tables in the investigation The If–Then Box. Follow the steps and use the sample data tables from Maze Puzzles to help you construct your data table for this investigation. You will become better at constructing data tables with each try.

Having a brainstorming session can be a lot of fun. During a brainstorming session, you can voice any idea that comes to your brain, no matter how crazy it might seem. You can give *any* idea that you think might provide a solution. Sometimes the ideas that seem far-fetched at first can lead to other ideas that really work. The main purpose of a brainstorming session is to create a storm in your brain so that you become very creative.

During a brainstorming session, you have permission to be as different in your thinking from others as you like. Your goal is to come up with as many different ideas as possible. After you finish your brainstorming session, you can come back to your task and use your ideas to solve a problem or to answer a question.

The process of brainstorming has a few guidelines, but not many. Before you begin your brainstorming session, decide how you will record your ideas. Then follow these guidelines:

1. State any idea about the topic that comes to your mind.

2. Record everyone's ideas. Don't judge whether the ideas are good or bad. Write them all down.

3. Keep thinking of ideas for at least 5 minutes. Continue for as long as you can or until your teacher tells you that your time is up.

4. If you can't think of a new idea, look at something that is already on the list. Try to add something to that idea or change the idea slightly.

5. If you are working in a group, take turns. Be sure that each person has a chance to suggest ideas. Remember that some people might need a little more time to think before suggesting something.

After you finish your brainstorming session, look at your list and decide which ideas might be better than others for solving the problem or answering the question. You should have a great list from which to choose!

In Unit 1 you learned about graphing. You learned about the parts of a graph in How To #2. In How To #3, you learned how to plot data to make bar graphs. In this How To, you will use what you learned in How To #2 and How To #3. Instead of showing the data with bars, though, you will use a line to connect the data points. When you connect the data points in this way, you make a **line graph.** A line graph helps you understand events that happen over time.

You and your classmates have just described what you saw on the maps in the investigation Patterns on the Earth. Probably a lot of what you described were the *patterns* that you observed on the maps. You could describe those patterns easily because you could see the data plotted on the actual maps, and you could point out the patterns to one another. But what if you had to explain the patterns to someone who wasn't in class and who didn't see the actual maps? There is another way to show some of the patterns you observed. You can use a graph.

In this How To, you will use the data from the Ages of Rocks map to show a pattern on a graph. Before you can make a graph, though, you must organize your data into a data table. (If you don't remember how to construct a data table, review How To #4.) To show the pattern from the Ages of Rocks map, you will need to display the ages of the rocks and their distance from the coast of South America. Your data table might look like the one in Figure H6.1.

To set up your graph, follow the same basic steps as you did in How To #3. Steps 1 through 5 will be the same. Plotting the data on your graph (step 6) will be different, however, because you will draw a line instead of bars. Let's go through the process one more time.

1. Review the data you have recorded.

   You will use data from the data table in Figure H6.1.

2. Draw the horizontal axis and the vertical axis for the graph.

3. Label each axis, using the headings in the data table.

   Because you want your line graph to help you see a pattern of something that happened over time, you should put the heading about time on the horizontal axis. In this case the heading for time is "Millions of Years." The other heading relates to distance, not time, so distance will be the label for the vertical axis.

4. Set up the number scales on each axis.

Use the data table to help you with this step. Notice that the numbers on the horizontal axis are from less than 1 million years to 190 million years and that each line represents 5 million years. The numbers on the vertical axis are from 0 kilometers to 2,200 kilometers, and each line represents 100 kilometers.

5. Give your graph a title.

6. Plot the data on your graph by doing the following:

- Read one row of data from the data table.

The first row of data tells you that the rocks that are 0 kilometers from the coast of South America (those rocks that are actually on or near the coast of South America) are 190 million years old.

- Find the number on the horizontal axis where that piece of data fits.

In this case the number you want on the horizontal axis is 190. Move your finger to that number.

- Move up from the number on the horizontal axis to the place on the vertical axis where that piece of data fits.

In this case you do not need to move your finger at all because the distance at 190 million years is 0 kilometers, which is on the horizontal axis (see Figure H6.3).

- Draw a dot, called a **data point,** at that place.

- Repeat the four parts of step 6 for every row of data in the data table.

For example, to plot the second data point, move your finger across the horizontal axis to the 65 million year mark and then up that line

Figure H6.1

This data table presents the data from the Ages of Rocks map.

Distance from the coast of South America in kilometers	Ages of rocks (in millions of years)
0	190
1000	65
1500	37
1900	22
2100	5
2200	<1

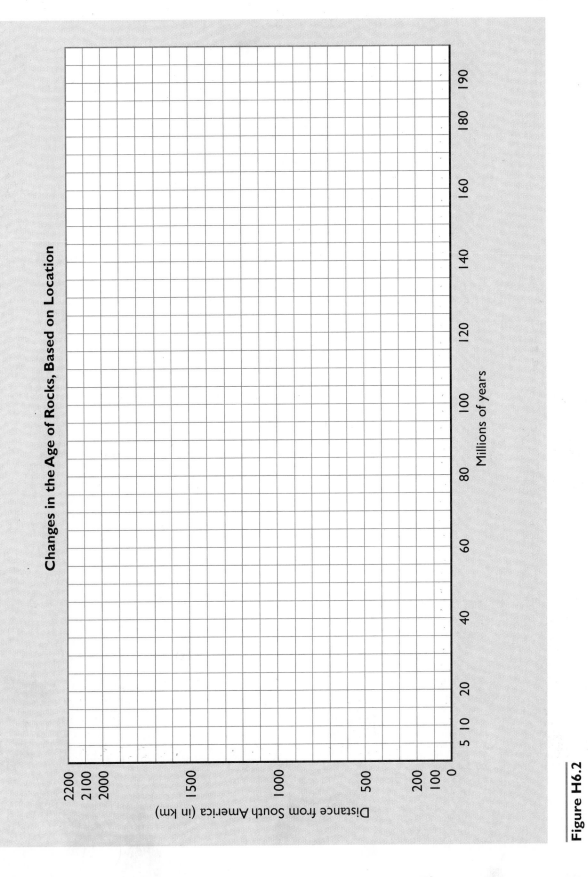

Changes in the Age of Rocks, Based on Location

Distance from South America (in km)

Millions of years

Figure H6.2

Now you are ready to plot data on your graph.

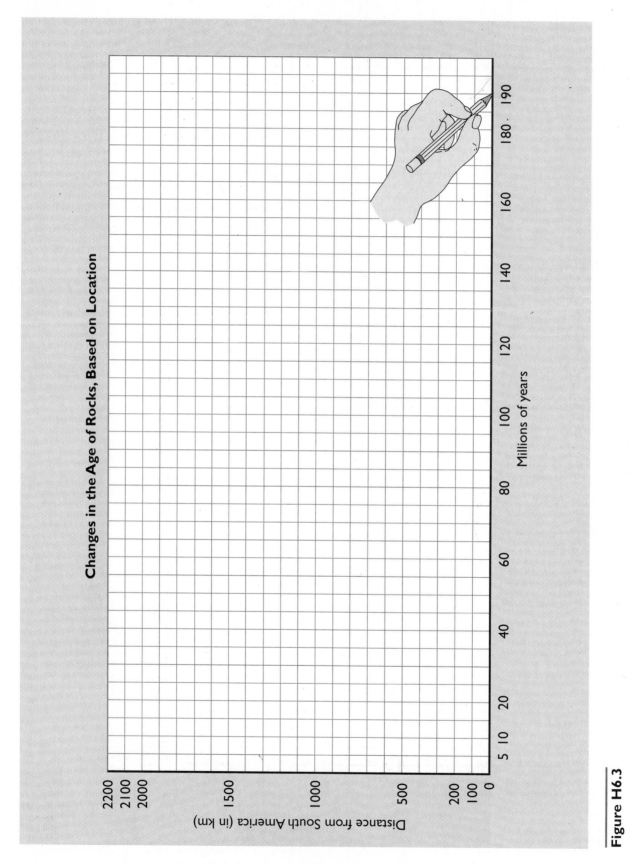

Changes in the Age of Rocks, Based on Location

Distance from South America (in km)

2200
2100
2000

1500

1000

500

200
100
0

5 10   20   40   60   80   100   120   140   160   180   190

Millions of years

Figure H6.3
This graph shows the first data point.

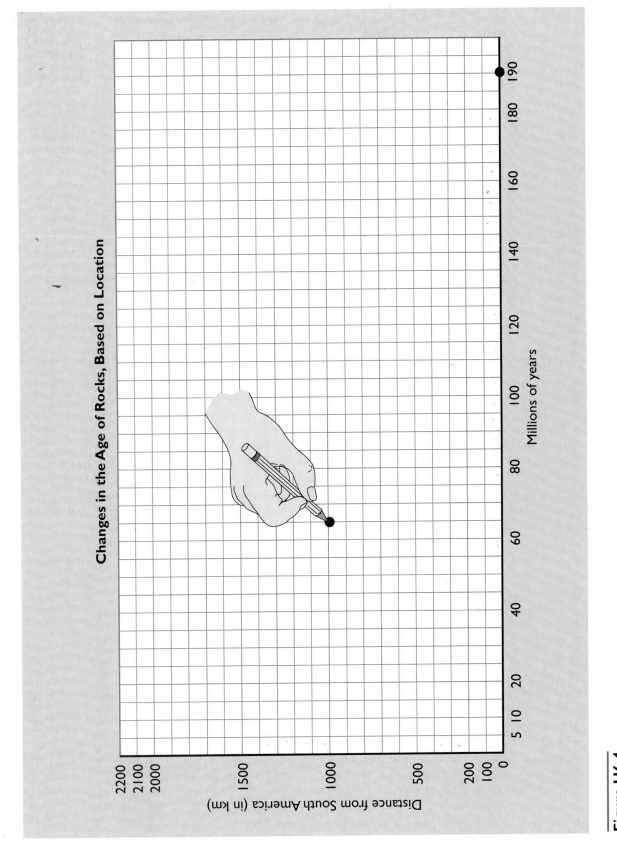

Changes in the Age of Rocks, Based on Location

Millions of years

Distance from South America (in km)

Figure H6.4
Can you begin to see a pattern?

to 1,000 kilometers on the vertical axis (see Figure H6.4). Make a dot for the second data point. Do the same for the rest of the data in the data table.

7. Draw a smooth line from left to right that connects all the data points.

You have completed a line graph that displays the data in the data table. Look at the pattern shown on the graph. Describe that pattern in terms of the age of the rocks and their distance from the coast of South America.

Questions:

1. As the rocks get older (in millions of years), do you move farther away from or closer to the coast of South America?

2. Where would you find the youngest rocks—near the coast of South America or toward the middle of the Atlantic Ocean?

3. What kind of pattern does this graph show?

A Bunsen burner is a type of heat source that burns fuel gas. This burner was designed in 1855 by a German scientist named Robert Bunsen. The Bunsen burner was a much more convenient and efficient heat source for a laboratory than either stoves or candles, which were the only other heat sources available then. The design of the Bunsen burner has changed little since Bunsen invented it, almost 150 years ago.

Part A—The Parts of a Bunsen Burner

The Bunsen burner has very few parts (see Figure H7.1). It has a fuel gas inlet, which is connected to the gas jet with rubber tubing. The burner has a movable ring near the base, which has openings called air ports. (The air ports look like small holes around the base of the Bunsen burner.) The air ports allow air to enter the barrel of the burner and mix with the fuel gas. Once ignited, the air and gas mix in the barrel to produce a flame at the top of the barrel.

You can change the size of the air port openings by rotating the ring. By changing the size of the air ports, you can control how much air enters the barrel of the burner. The amount of air entering the Bunsen burner is important because fuel gas cannot burn without air. If not enough air is mixing with the gas, then the gas will not burn completely, and the flame won't get very hot.

Part B—Safety Cautions When Using a Bunsen Burner

Always be careful when lighting and using a Bunsen burner. Wear eye protection at all times. Be sure you do not put your face near the top of the barrel at any time. Tie back long hair and remove scarves, ties, or jewelry that might hang down into the flame. Remember that the flame of a Bunsen burner burns very hot; do not play around during an investigation that requires the use of a Bunsen burner or any other heat source.

Strike the match *before* you turn on the gas jet. Place the match to the side of the barrel, just below the top. (If you hold the match directly over the barrel, the gas jet will extinguish the flame.) As soon as you hear the gas, light the burner. The gas should not be on longer than a few seconds before you light it. If you have trouble lighting the burner, turn off the gas and try again. *Do not keep the gas flowing while you wait to light a new match.* If you fail to light the burner after a few attempts, ask your teacher for help.

Figure H7.1

A Bunsen burner has few moving parts and is easy to use. A Bunsen burner burns with an open flame, though, so you must always be careful when you use one.

Barrel

Moveable ring

Air ports

Gas inlet

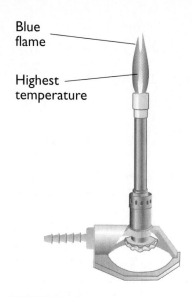

Blue flame

Highest temperature

Figure H7.2

When you have adjusted your Bunsen burner properly, you should see a flame that looks like this.

After you have finished with the burner, be sure you turn off the gas jet completely so that no gas escapes into the room.

Part C—How to Light and Adjust a Bunsen Burner

Follow these steps in lighting and adjusting a Bunsen burner.

1. Put on your safety goggles.
2. Turn the ring near the base of the burner so that the air port openings are almost closed.

    The air port openings should be open just slightly so that a little air flows through them.

3. Strike a match. (Use a long wooden one.)
4. Turn on the gas jet.
5. Hold the match just to the side of the barrel of the burner.

    The burner should light and probably will burn with an orange flame. The orange flame is a sign that the gas in the burner is not burning completely.

6. Adjust the flow of air and the flow of gas until you see a blue flame.

    When the burner is burning properly, you will see a blue flame that has two distinct parts (see Figure H7.2). To get the blue flame, slowly open the air ports. This will add more air to the fuel. You might need to adjust the flow of gas from the gas jet as well. Keep adjusting the air port openings and the gas flow until you get a steady, blue flame.

    The gas should burn quietly. If you hear a roaring sound, then too much gas probably is coming into the barrel. You should reduce the flow of gas and possibly close the air port openings a little, too.

7. After you have finished with the Bunsen burner, turn off the gas jet completely and close the air ports.

    The flame should go out quickly.

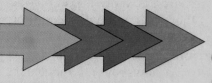
So you are going to do a research project. It might seem like a pretty big deal, but it's not if you take it one step at a time. Actually conducting a research project *will* take more time than a simple homework assignment, but it also should be more interesting. There are two keys to conducting good research for a report or a presentation: (1) Choose a topic that interests you, and (2) get organized. If you go about your research in an organized manner from the beginning, then it will be a snap to put your presentation or report together. This How To has some tips that will help you get organized.

You might be wondering what research is. How is conducting research different from completing other assignments? According to the *American Heritage Dictionary,* to research a topic means to study the topic thoroughly. Therefore conducting research is different from just reading one article in a newspaper, a magazine, an encyclopedia, or a textbook. When you conduct a research project, you gather as much information as you can about the topic you are researching. Then you put the information together in an organized way that will make sense to someone who listens to your presentation or reads your report.

Tips for Conducting a Research Project

Part A—Choosing Your Topic

Tip #1: List several topics that interest you.

Usually you can't just think up your own topics. Your teacher or your textbook probably will limit your choices to a general area of research. In this case your assignment is to research a topic that is related in some way to plate tectonics.

To get started, list at least five possible topics that are related to the general subject of your research—plate tectonics, in this case. Then rank the topics and circle your first and second choices. (You always should have two choices, because you might not be able to find enough information about your first choice.)

Stop now and make your list of five or more possible choices. You may write your list on BLM HT8.1.

Part B—Getting Organized

Tip #2: Think about your topic before you read anything.

You might ask yourself these questions about the topic you have chosen:

- What do I already know about the topic?
- What books or magazines have I seen that might have some information about the topic?
- What would I like to know about the topic?
- What *about the topic* would I like to share with my classmates?

If you organize your thoughts first, then it will be easier to organize the information you find.

Let's say you chose the topic of geysers (GUY zerz) for your research project. After thinking about the topic of geysers, you decide you want to answer the following questions:

- How are geysers formed?
- What do geysers look like?
- What makes geysers shoot up into the air?
- What are the names of some famous geysers around the world?

You might organize your research by writing each of those questions at the top of one sheet of paper. Then as you find information that answers one of those questions, you can write the information on that sheet of paper. That way, you may organize your notes before you even start. (For your notes you may use note cards instead of sheets of paper. If you do use note cards, you probably will need more than one card for each main idea or question.)

As you read, stay open to new ideas. You might think of new questions about your topic that you would like to answer. Start a new sheet of paper or a new set of note cards for each main idea that you find interesting.

Stop now and write some things you already know about the topic and some of the questions you would like to answer. You may write your ideas and questions on BLM HT8.1.

Part C—Finding Information

Tip #3: Look in more than one source for information.

Several different resources are designed especially to help you locate information:

- *Reader's Guide to Periodical Literature*
- on-line (computerized) catalogs or databases
- the card catalog

But how can you find the right sources that will have information about your topic? First you need to know enough about your topic to identify a few **key words.** Key words are important words related to the topic you want to study. Sometimes you also can come up with key words by brainstorming a list either by yourself or with a partner. Ask yourself:

- What do I already know about my topic?
- What words do I associate with my topic?

For example, let's use the topic of geysers again. To get started, you might look up the word "geyser" in an encyclopedia or in a dictionary. There, you might find words, such as "hot springs," "geothermal energy," and "hot spots." Then, you remember reading about a geyser named "Old Faithful." The words inside the quotation marks are all possible key words because they tell you something about geysers.

You will need to have some key words in mind before you can use effectively the card catalog, an on-line data base, or the *Reader's Guide to Periodical Literature.* If you have not used those resources before, ask your media specialist, librarian, teacher, or a friend for help.

Stop now and list the resources you will use to get started. Then list a few key words that will help you find out more about your topic. You may add to your list of key words as you do your research. Write your lists in the appropriate spaces on BLM HT8.1.

After you have learned to use the *Reader's Guide to Periodical Literature,* the on-line data bases, or the card catalog, these resources will lead you to specific articles in newspapers or magazines or to books about your topic.

In addition to these sources of information, the following sources of information might be helpful:

- science dictionaries
- pamphlets
- audiovisual materials such as films, filmstrips, videotapes, videodiscs, slides, audiotapes, and television specials or documentaries
- people who are experts on the topic or have experience related to the topic
- encyclopedias

Part D—Taking Notes

Tip #4: Take notes in your own words.

Sometimes you might think it is easier just to read something in an encyclopedia and copy whatever it says. That type of library research has two problems. First, copying what someone else has written and using it as your own work is called **plagiarism** (PLAY-jer-ism), and plagiarism is illegal. Second, your report or presentation will be much more interesting if you put the information together in your own words. You might add something humorous that you found out, or you might weave your information into an exciting adventure story about a geologist who discovered a new geyser. Copying from encyclopedias probably will make your report uninteresting to your classmates, too.

Sometimes, though, you might want to write down the exact words someone wrote or said. That is okay, as long as you use those words as a **quote.** Quotes can make your presentation or report more interesting, because a quote can add humor or authority to your report. When you use a quote in your presentation or report, you *must* use the *exact words* and tell or write down who said the words and where you found the quote. Using a quote is not plagiarism because you are giving credit to the person who said or wrote those words. You are not trying to say that those are *your* words. Be sure each quote you use is not more than about 50 words long. You should use quotes to "spice up" your report, not to *be* your report.

You might want to keep track of the direct quotes you find by writing them on BLM HT8.2.

Part E—Making a Reference List

Tip #5: Write complete information about all of the reference materials you use.

Before you put away a book, magazine, or pamphlet from which you took notes, write down the following information:

- the title of the book or magazine
- the title of any specific article you read from a book, magazine, or pamphlet
- the author or authors (if any)
- the publisher
- the city where the book or magazine was published
- the copyright date of a book
- the volume if it is an encyclopedia, and the date if it is a magazine or newspaper
- the page numbers you read

Your teacher probably has a certain way she or he would like you to list your references. Be sure you know how you should list your references before you put those references away. It is time consuming to go back and find all your sources again if you need more complete information.

As you use reference materials for your research project, write down the necessary information about each reference on BLM HT8.2. That way, you will have your reference list almost ready when you finish taking notes.

Part F—Organizing Your Notes

Tip #6: Separate your notes into different main ideas.

After you have gathered all the information you need, look through your notes. Decide what information you want to include in your presentation or report and what information might be unnecessary. You don't have to use all of the information that you gathered, just because you have it. It is always better to have more information than you need.

Then if you have not already done so with separate papers or note cards, divide your notes into different main ideas. Go back to Part B—Getting Organized, and review the main ideas you wrote on BLM HT8.2. Use those main ideas and others that you chose as you gathered information.

Finally organize your main ideas into a logical order. What do you want to talk or write about first, second, and third? Write down your main ideas in order on BLM HT8.3. Then decide what information would make the best introduction and what information would make the best ending. Write your ideas for an introduction and for an ending in the appropriate boxes on BLM HT8.3. If you know how, you might put your notes into an outline form.

Before you go to Part G, decide what information would be best to show on a poster or with a model or diagram. Separate that information from the rest of your notes and write down your ideas on BLM HT8.3.

Part G—Writing Your Report or Planning Your Presentation

Tip #7: Don't worry about every detail in your first draft.

When you conduct a research project, you will write more than one draft of your report. Don't try to get all your ideas down exactly the way you want them on your first draft. The most important thing to decide is the order of your main ideas. (Review what you wrote on BLM HT8.3.) Then you can decide which details best support those main ideas. After you complete the first draft of your report or presentation, you can go back through it and add details, check the flow of your ideas, and be sure you haven't left out anything important. You now are ready to write a revision of your paper or presentation.

Begin with a strong introduction that will make your classmates interested in your presentation. Be sure to state the most important idea in your introduction.

Organize your information into complete sentences that flow in a logical order. Move smoothly from one main idea to the next,

Wow! It's really finished! I'm really proud of my research report.

supporting all your information with examples and quotations from your research. Be sure to support your opinions with information, quotes, and examples from your research. Also, think about what information would be best to present on a poster, in a diagram, or as a demonstration.

Decide how you will end your presentation or report. Your conclusion should tie all your information together and summarize your main ideas.

Use the information you have organized on BLMs HT8.2 and HT8.3 to help you through this process.

Part H—Congratulating Yourself on a Job Well Done

Conducting a research project might not be the easiest thing you have ever done, but it can be rewarding. Often you will find interesting information that you never knew before. You probably will remember it longer, too, because you really had to think about how all the information fit together. When you have finished your report, congratulate yourself on a job well done.

Averaging numbers can be a useful skill when you want to find out what *usually* happens or what is likely to happen in a particular situation. For example let's say that your family plans to take a vacation in August and that you want to travel to a state where tornadoes are unlikely to occur. Before you could decide where to go, you would need to know how many tornadoes usually occur in different states in August. Look at the data in Figure H9.1.

Part A—Finding an Average

When you look at these data, you will see that the number of tornadoes that occurred each year in each state was not always the same. In some years quite a few tornadoes occurred in August, and in other years few or no tornadoes occurred. How could you predict how many tornadoes might occur this year in August in each state? The best way to make your prediction would be to find the **average** of the numbers of tornadoes that occurred each year in August in each of the states.

You need to follow only two steps to find an average.

1. Find the **sum** (the total) by adding all the numbers together.

   The total number, or sum, of tornadoes in August in Nebraska from 1985 through 1989: 6 + 3 + 4 + 3 + 4 = 20.

2. Divide the sum by the number of data points you have.

   In this case you would divide the sum by 5 because you have data for 5 years, each year from 1985 through 1989.

   In Nebraska the total number of tornadoes in August from 1985 through 1989 was 20. You have 5 years' worth of data, or 5 data points, so divide 20 by 5:

   20 tornadoes in August ÷ 5 years = 4 tornadoes per year in August

Now you know that the average number of tornadoes that occurred in Nebraska in August from 1985 through 1989 was four. This means that you might expect four tornadoes to occur in Nebraska each year in August. As you can see from the actual data, the number of tornadoes was above the average some years and below the average other years. The average simply tells you the number of tornadoes that *usually* occur in August in Nebraska.

Year	Number of Tornadoes in August in Nebraska
1985	6
1986	3
1987	4
1988	3
1989	4

Year	Number of Tornadoes in August in Florida
1985	9
1986	9
1987	6
1988	2
1989	4

Year	Number of Tornadoes in August in Wisconsin
1985	5
1986	0
1987	5
1988	6
1989	0

Figure H9.1

Which state is likely to have the highest number of tornadoes in August?

Part B—More Practice Finding Averages

Look at the data for the other two states. Answer these questions *before* you find the average number of tornadoes for those states.

Questions:

1. How will the average number of tornadoes in August in Florida compare with the average number of tornadoes in August in Nebraska? (Will the average in Florida be higher or lower?) Explain your answer.

2. How will the average number of tornadoes in August in Wisconsin compare with the average number of tornadoes in August in either Nebraska or Florida? (Will the average in Wisconsin be higher or lower?) Explain your answer.

When you look at the data for Wisconsin, you will notice that *no* tornadoes were reported in August in 1986 or in 1989. How will you find the average when some of the data points are zero? You still follow the same steps.

1. Find the **sum** by adding all the numbers together.

$5 + 0 + 5 + 6 + 0 = 16$

2. Divide the sum by the number of data points you have.

Even though some of the numbers are zero, you still have one data point for each year for 5 years. Some of those data points just happen to be zero. You still divide the sum by 5 because you added 5 numbers (5 data points) together to get the sum.

16 tornadoes ÷ 5 years = 3.2 tornadoes per year

(You also can write 3.2 tornadoes per year as 3 tornadoes per year with a remainder of 2/10 or 1/5.)

As this example shows, sometimes the numbers do not divide evenly, and you end up with a whole-number average with some amount left over. In those cases you will use a skill you already have: the skill of rounding off numbers. (See How To #1 if you do not remember how to round off numbers.) It is not possible for there to be 2/10ths of a tornado; there either is a tornado or there isn't one. So, in the state of Wisconsin, you would say that the average number of tornadoes in August is three.

Before going on to Part C, follow the steps and find the average number of tornadoes that occurred in August in Florida.

Part C—Using Averages to Make Predictions

Does the average tell you *exactly* how many tornadoes *will* occur in August in each of these states this year? No, averages do not give you exact numbers. They simply tell you what is likely to happen based on the data you have.

Averages can be very useful numbers, especially if you want to make predictions based on what usually happens. For example, if you are a baseball pitcher, you might like to know which batters are more likely to hit the ball when they are up to bat. How would you find out? You would use each batter's batting average, which is based on the total number of hits the batter got, divided by the number of times that batter was at bat. Those players who have a high batting average are more likely to get a hit when they come up to bat than those players who have low batting averages.

Weather forecasters also use averages when they discuss the weather. Sometimes temperatures in the summer might be unusually high, and the weather forecaster might say, "The temperatures continue to be above average for this time of year." This statement means that the temperatures are above those that *usually* occur at that time of year.

Scientists called climatologists also make use of average temperatures. They are studying average temperatures over time to decide whether the earth is experiencing something called "global warming." Those scientists are trying to find out whether temperatures around the world are increasing over time or whether they are staying basically the same. To make their predictions, scientists use recorded average temperatures from many places around the world over many years.

Listen carefully to news reports and weather forecasts on television. How many times do you hear the newscasters or weather forecasters talk about averages? The next time you hear someone use the term "average," you will know exactly what he or she means.

What does 20% mean? Are Marie's chances of winning good or poor? The symbol % stands for "**percent.**" The word "percent" means "out of 100." So 20% reads "20 percent," or "20 out of 100."

To make percentages easier to understand, let's use a diagram. Suppose 100 total chances were available in Marie's sweepstakes. We will use a grid of 100 squares to represent the 100 total chances. Thus on this grid each square represents 1 chance (see Figure H10.1).

This certificate says that I have a 20 percent chance of winning a sweepstakes. Am I likely to win? I wonder what 20 percent really means?

Figure H10.1

This grid shows all the possible chances in Marie's sweepstakes. Each square represents one chance.

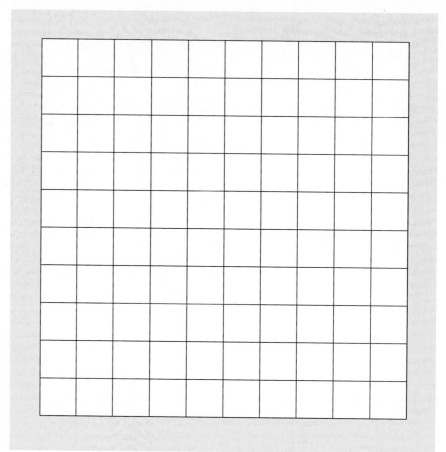

The certificate suggests that Marie has a 20 percent chance of winning. If we say that each red square represents one chance of Marie winning, how many squares of the grid would be red? (See Figure H10.2.)

That's right. If 20 percent means 20 out of 100, then 20 of the 100 squares would be colored red.

Look at the grid in Figure H10.2 again. What do the 80 white squares stand for? If the 20 red squares stand for the number of chances Marie has of *winning,* then the 80 white squares stand for the number of chances Marie has of *not winning* (100 total chances − 20 chances of winning = 80 chances of *not* winning.) So are Marie's chances of winning good or poor? Use the data shown on the grid to answer the question. Be prepared to support your answer.

Often people use percentages to represent **probability.** (Remember that probability can tell you how *likely* something is to happen, but it cannot tell you that something *will* happen for sure.) Let's go back to Marie's sweepstakes. This time, we will use 100 plastic pieces to represent the total chances of winning. We will show Marie's 20 percent chance of winning by including 20 red pieces in the 100 total pieces. The other 80 pieces will be white. The white pieces show the chances Marie has of *not* winning.

Imagine that you put these 100 plastic pieces into a bag and that, without looking into the bag, you drew out 1 plastic piece. Do you think the piece is likely to be a white piece or a red piece? Because the majority of the pieces are white, it is more likely that you would draw out a white piece. But you *could* draw out a red piece, couldn't you? It is just not as likely that you would draw a red piece because there are so few of them. So once again this shows that Marie's chances of winning the sweepstakes are not very good.

Suppose that Marie's sweepstakes certificate had said that she had a 60 percent chance of winning. Look at the change in the plastic pieces when 60 percent of them are red. If the red pieces represent the chances of winning, decide whether you think Figure H10.4 accurately shows a 60 percent chance of winning.

Figure H10.2

This grid shows the 20 chances of winning as 20 red squares.

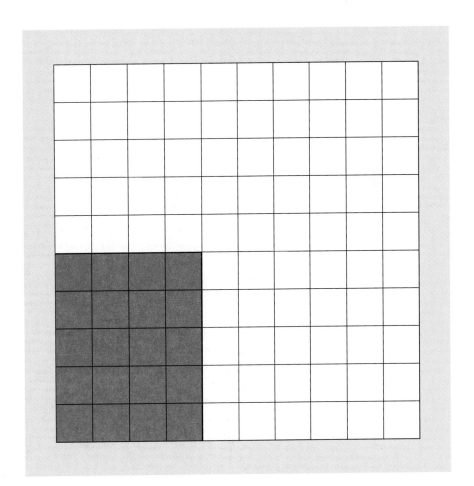

Figure H10.3

Twenty percent, or 20 out of 100, of these pieces are red.

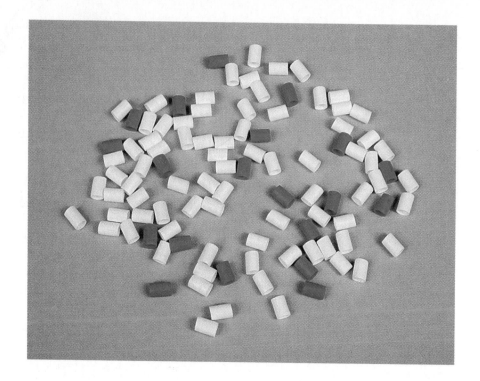

Percentages usually aren't difficult to understand when you have exactly 100 of something. But suppose you have 50 plastic pieces and that 20 percent of them are red. How many of the 50 pieces would be red? Look back at the photograph of the 100 pieces when 20 percent were red. Then imagine that someone divided the

Figure H10.4

Would Marie be more likely to win if she had a 60 percent chance of winning?

Figure H10.5

This photograph shows the 100 plastic pieces divided into two equal piles.

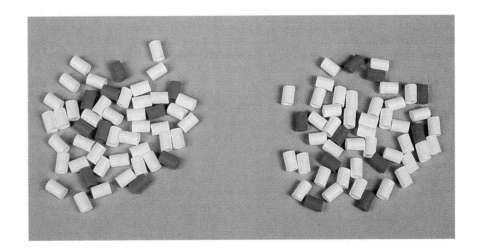

100 pieces into two *equal* piles. To be exactly equal, each pile would have exactly the same number of white pieces and the same number of red pieces (see Figure H10.5).

Now you have 50 plastic pieces in each pile. Count the number of red pieces in each pile. You should count 10 red pieces in each pile. This means that 10 red pieces out of 50 total pieces is the same as 20 red pieces out of 100 total pieces. The portion of pieces that are red, compared to the total number of pieces is the same. You can look at this in terms of fractions, too:

$$\frac{20 \text{ red pieces}}{100 \text{ total pieces}} = \frac{10 \text{ red pieces}}{50 \text{ total pieces}}$$

Figure H10.6

How many red pieces are in each pile if 20 percent of the pieces are red?

Suppose you have a total of only 10 plastic pieces and that 20 percent of those are red. How many of the plastic pieces in each pile

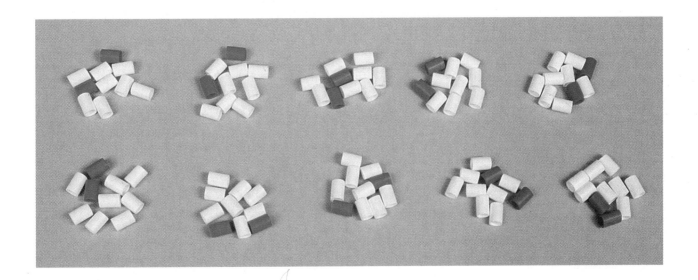

of 10 would be red? To find out, divide the original pile of 100 pieces into 10 equal piles. (To make the piles equal, each pile of 10 must have the same number of white pieces and the same number of red pieces.) If you make the piles equal, you should count 8 white pieces and 2 red pieces in each pile of 10.

Look back at the photographs. Do you see that 2 red pieces out of 10 is the same distribution as 10 red pieces out of 50 and 20 red pieces out of 100? You can look at the relationship between the numbers in terms of fractions:

$$\frac{20 \text{ red pieces}}{100 \text{ total pieces}} = \frac{10 \text{ red pieces}}{50 \text{ total pieces}} = \frac{2 \text{ red pieces}}{10 \text{ total pieces}}$$

Now try to solve these problems:

1. You have 10 plastic pieces and 40 percent of them are red. How many pieces are red and how many are white?

2. You have 10 plastic pieces and 60 percent of them are red. How many pieces are red and how many are white?

Glossary

air masses: Air masses are areas with air of similar temperature.

average: To calculate an average, find the *sum* by adding all the numbers together. Then divide by the number of data points or items that you have. For example a store owner might sell 10 bags of carrots in 2 days. The bags of carrots are the data, and the owner collected data for 2 days. The owner sold an *average* of 10 ÷ 2, or 5 bags of carrots per day.

axis: An axis is the horizontal or vertical line in a graph on which we write numbers or labels. Most graphs have both a horizontal and a vertical axis, and the place where these two lines meet is where the graph begins. (The plural of *axis* is *axes*.)

bar graph: A bar graph uses bars to illustrate the results of a survey or an experiment.

benefits: Benefits are the positive effects of a decision.

buoyant force: Objects float because of buoyant force. The force pushing down is less than the force of the water (or other material) pushing up.

cause and effect: Cause and effect is a pattern in which one or more factors make something happen.

centimeter (cm): A centimeter is a small unit of length equal to one-hundredth of a meter. A centimeter is slightly less than one-half of an inch.

combustion: Combustion is when something ignites or bursts into flame.

compass: A compass is an instrument that shows which direction a person is facing—north, south, east, or west.

compost: Compost is usually plant material (such as leftover food and shredded paper) that breaks down into nutrient-rich material. This nutrient-rich material also occurs in soil (when leaves and grasses rot). Compost can be used to enrich soil.

condensation: Condensation occurs when water particles in the air cool and move together. Water vapor then becomes liquid water.

continental drift: Continental drift is Alfred Wegener's geologic theory. He proposed that because the shapes of the continents fit together like a puzzle, it was likely that the continents had been together at one time and later moved to their present locations.

controlled experiment: An experiment in which only one factor is changed and all the other factors remain the same is called a controlled experiment. If experimenters want to see whether one factor is affecting a pattern, they must do a controlled experiment (a fair test).

convection cell: A convection cell is a pattern in which cold fluid (for example air or water) sinks and warm fluid rises, creating a somewhat circular pattern.

core: The core is the very innermost part of the earth. The core is made up of a liquid outer core and a solid inner core.

Coriolis effect: The Coriolis effect is when something moving in a straight line appears to curve because it is moving above a spinning surface.

correlation: A correlation is a pattern in which two or more things are related or connected to each other. For example in humans the length of a person's forearm is about the same as the length of his or her foot, and this is a correlation.

costs: Costs are the harmful or negative effects of a decision.

crust: The crust is the outer shell of the earth, which consists of mostly hard rock.

cycle: A cycle is a repeating pattern, such as the repeating pattern of daylight and darkness.

data: Data can be things such as measurements, numbers, observations, dates, or time. *Data* is the plural of the word *datum*. Because most often we are working with more than one piece of information, we say "data."

data point: A data point is a point on a graph that indicates the observed results.

data table: A data table is a chart that helps a person keep track of observations.

drought: Drought is a time when no rain or snow falls in an area that needs water. A drought may last for months or years.

evaporation: Evaporation occurs when water particles move from a liquid into the air.

evidence: Evidence is information that you obtain using your senses: you can see, hear, smell, taste, or feel evidence.

explanation: An explanation is more than just a yes-or-no answer to a question. When someone proposes an explanation, he or she can justify why it is probably accurate.

eye: The eye is the calm area in the center of a hurricane.

factors: Factors are things that influence patterns, so scientists consider factors when they set up experiments.

faults: Faults form when forces within the earth push rock to the point of breaking. These forces may push a broken block of rock up, down, or sideways. Movement along a fault can push one type of rock next to another type of rock.

floodplains: Floodplains are flat areas next to rivers. This area is the first place to be flooded if a river overflows its banks.

folding: Folding occurs when rock layers bend but do not break.

fossils (FAH suhls)**:** Fossils are the hardened remains or traces of animals or plants that lived long ago.

graph paper: Graph paper is special paper with evenly spaced lines. The lines run from side to side and from top to bottom, forming a grid.

groundwater: Water that moves through the soil and underground is called groundwater.

horizontal axis: The horizontal axis is the line in a graph that runs from left to right.

hot spots: Some places on the earth have a lot of volcanic activity and yet are not near a plate boundary. These are called hot spots. Two examples of places affected by hot spots are the Hawaiian Islands and Yellowstone National Park.

hurricane: A hurricane is a large storm that forms over warm oceans near the equator and that causes damage with floods and strong winds when it moves onto land.

incinerators: Garbage is burned in furnaces called incinerators.

key words: Key words are the important words that relate to the topic that a person chooses to study. For example, if a person were studying about patterns of grocery shopping, *food* would be an important key word.

landfill: A landfill is most often a large pit or hole where people put garbage. The landfill can become so large that it is a mound. In some landfills, called sanitary landfills, layers of garbage are covered with dirt each day to keep the garbage from blowing away.

line graph: When a line connects or is drawn near the data points on a graph, a line graph is formed.

magma: Magma is hot matter (molten rock) beneath the earth's surface. When it reaches the earth's surface, it is called lava.

mantle: The layer beneath the earth's crust that is more like gelatin than like hard rock is called the mantle. The mantle is softer than the crust because of the heat and pressure inside the earth.

millimeter (mm): A millimeter is a very small unit of length equal to one-thousandth of a meter. A millimeter is smaller than one-eighth of an inch. Millimeter marks are between centimeter marks on a metric ruler.

mountains: Mountains are high places on the earth's crust. They form by volcanic eruption, by folding when plates collide, or by movement along some types of faults.

number scale: A sequence of numbers plotted along a line is called a number scale. The number scale may be vertical or horizontal. If the number scale is on a graph, the numbers usually are read from left to right on the horizontal axis and from bottom to top on the vertical axis.

pattern: A pattern is a collection of things or events that repeat themselves.

pendulum (PEN dyew lum): A pendulum is a weight that hangs from a string.

phase: The word *phase* comes from the Greek word *phaino*, which means to "appear or to bring to light." In this book we talk about the phases of the moon as a pattern that we can see from the earth. A phase is a change that is "brought to light," or acknowledged.

plagiarism (PLAY jer ism): Plagiarism consists of copying what someone else has written and using it as your own work. Plagiarism is illegal.

plates: The continents are part of plates of rock that make up the entire surface of the earth.

predictions: A prediction is a statement about the future, based on information. A prediction differs from a guess because a guess is not based on information.

probability: Statements of probability indicate how *likely* an event is to occur, but they do not usually indicate whether the event *actually* will happen.

quality: The quality of something refers to its degree of excellence or character.

quantity: A quantity is the measured amount of something.

quantity and quality: Quantity and quality are two factors to consider when gathering information. To make accurate predictions based on information, you need enough of the best kind of information.

quote: A quote contains the exact words someone wrote or said, with an acknowledgment of the source.

recycling: Recycling means that materials repeat a cycle or that people use them more than once.

ridges: Ridges are the places where two plates are moving apart. On earth, most ridges are towering mountain chains on the ocean floor that are as high as 2,400 meters (7,880 feet) and that encircle the earth like seams on a baseball.

round off: To change a mixed number (e.g., 2_ or 2.5) to the nearest whole number (for example, 3) is called rounding off. The basic rules for rounding off numbers are: any fraction less than 0.5 rounds *down,* and any fraction from 0.5 and higher rounds *up.*

San Andreas fault: One example of a transform fault is the San Andreas fault in southern California. When the Pacific and the North American Plates move along this fault, earthquakes occur.

scientific explanation: A scientific explanation is more than just a yes-or-no answer to a question. A scientific explanation explains the *cause* of a pattern and is based on information. With a scientific explanation, a scientist can explain *why* he or she feels that a particular idea is probably accurate.

sum: The total of a group of numbers added together is called a sum.

technology: Technology is a process of designing and building things that solve people's problems. Technology can be very simple, like a can or a toothpick, or more complex, like a car or a computer.

test: A test is a series of questions or problems designed to prove whether an explanation is correct.

theory of plate tectonics: This theory proposes that the earth's surface is covered by a set of interlocking plates and that these plates move slowly across the earth's surface.

tornado: A tornado is a storm in which winds of incredibly high speeds move in a funnel cloud. Tornadoes usually occur with severe thunderstorms.

toxic or **hazardous waste:** Trash that is potentially harmful to people is called toxic or hazardous waste.

transform faults: Transform faults occur when two plates slide past each other. When this movement occurs, it can produce earthquakes.

trenches: Trenches are the deepest parts of the ocean. At a trench one edge of a plate sinks beneath another plate.

trend: A trend is a type of pattern in which change occurs in a particular direction. One example would be that the use of a particular item, such as umbrellas, continues to increase (goes up).

vertical axis: The line in a graph that runs up and down is called the vertical axis.

water cycle: The water cycle is the pattern of water movement on the earth, in which water particles move from liquid to vapor and back to liquid again. The water cycle occurs as water moves from lakes or oceans into the clouds, falls as rain or snow, moves through the ground and into creeks and rivers, and rises back into the clouds.

wind: Wind is air that is moving horizontally in a convection cell.

(Board Members continued from p. ii.)

Tracy Posnanski, *University of Wisconsin-Milwaukee, Milwaukee, Wisconsin*

Douglas Reid, *Southridge Middle School, Fontana, California*

Rochelle Rubin, *Instructional Materials Center, Waterford, Michigan*

Charlotte Schartz, *Kingman Elementary School, Kingman, Kansas*

M. Gail Shroyer, *Kansas State University, Manhattan, Kansas*

Elayne Shulman, *Classroom Consortia Media, Metuchen, New York*

Barbara Spector, *University of South Florida, Tampa, Florida*

John Staver, *Kansas State University, Manhattan, Kansas*

John Swaim, *University of Northern Colorado, Greeley, Colorado*

Robert Tinker, *Technical Education Research Centers, Cambridge, Massachusetts*

David Trowbridge, *University of Washington, Seattle, Washington*

Project Advisors and Consultants

William D. Gillan, *IBM, Boca Raton, Florida* (Corporate Advisor for Design Study)

Martin Guttmann, *IBM, Boca Raton, Florida* (Corporate Advisor for Design Study)

Ann Haley-Oliphant, *Mainville, Ohio* (Contributing Author)

Norris Harms, *Arvada, Colorado* (Evaluation)

A. W. Harton, *IBM, Atlanta, Georgia* (Corporate Advisor for Design Study)

James McClurg, *University of Wyoming, Laramie, Wyoming* (Curriculum Development)

Ann Primm, *Knoxville, Tennessee* (Contributing Author)

James R. Robinson, *Boulder, Colorado* (History)

M. Gail Shroyer, *Kansas State University, Manhattan, Kansas* (Implementation)

Dave Somers, *Colorado Springs, Colorado* (Editor)

Terry G. Switzer, *Fort Collins, Colorado* (Contributing Author)

Luise Woelflein, *Washington, DC* (Contributing Author)

Field-Test Sites

Primary Site Centers and Affiliated Schools

California
Almeria Middle School, Fontana, California, 1990–91

Southridge Middle School, Fontana, California, 1990–92

Coordinated by Herbert Brunkhorst (Site Coordinator) and Carol Cyr (Graduate Assistant 1990–91) and Cynthia Peterson (Graduate Assistant, 1991–92) based at California State University, San Bernardino, California.

Colorado
Carmel Middle School, Colorado Springs, Colorado, 1990–92

Challenger Middle School, Colorado Springs, Colorado, 1990–92

Colegio Los Nogales, Bogota, Colombia, South America, 1991–92

The Colorado Springs School, Colorado Springs, Colorado, 1990–92

Desert School, Rock Springs, Wyoming, 1991–92

Eagleview Middle School, Colorado Springs, Colorado, 1990–91

East Junior High School, Rock Springs, Wyoming, 1991–92

Gorman Middle School, Colorado Springs, Colorado, 1990–92

Panorama Middle School, Colorado Springs, Colorado, 1990–92

Smiley Middle School, Denver, Colorado, 1991–92

Timberview Middle School, Colorado Springs, Colorado, 1990–91

White Mountain Junior High School, Rock Springs, Wyoming, 1991–92

Coordinated by BSCS staff based in Colorado Springs, Colorado.

Florida
Clearwater Comprehensive School, Clearwater, Florida, 1990–91

Harllee Middle School, Bradenton, Florida, 1991–92

Lincoln Middle School, Palmetto, Florida, 1991–92

16th Street Middle School, St. Petersburg, Florida, 1990–92

Southside Fundamental School, St. Petersburg, Florida, 1990–92

W. D. Sugg Middle School, Bradenton, Florida, 1990–92

Coordinated by Barbara Spector (Site Coordinator) and Merton Glass (Graduate Assistant) based at University of South Florida, Tampa, Florida.

Kansas
Chapman Middle School, Chapman, Kansas, 1990–92

Dawes Junior High School, Lincoln, Nebraska, 1991–92

East Junior High School, Lincoln, Nebraska, 1991–92

Fort Riley Middle School, Fort Riley, Kansas, 1990–92

Kingman Middle School, Kingman, Kansas, 1990–92

Murdock Elementary School, Kingman, Kansas, 1990–92

Norwich High School, Kingman, Kansas, 1990–92

Norwich Junior High School, Kingman, Kansas 1990–92

Pound Junior High School, Lincoln, Nebraska, 1991–92

Coordinated by John Staver (Site Coordinator) and Randall Backe (Graduate Assistant, 1989–91) and Ronald Krestan (Graduate Assistant, 1991–92) based at Kansas State University, Manhattan, Kansas.

New York
Roy W. Brown Middle School, Bergenfield, New Jersey, 1991–92

Longwood Junior and Senior High School, Middle Island, New York, 1990–91

Longwood Middle School, Middle Island, New York, 1990–91

Mount Sinai Middle School, Mount Sinai, New York, 1991–92

Shoreham-Wading River Middle School, Shoreham, New York, 1990–91

Southampton Intermediate School, Southampton, New York, 1991–92

Tremont School, Mount Desert, Maine, 1990–91

Coordinated by Thomas Liao (Site Coordinator) and Rita Patel-Eng (Graduate Assistant, 1989–91) and Cynthia Anderson (Graduate Assistant, 1991–92) based at State University of New York, Stony Brook, New York.

Ohio
Dater Junior High, Cincinnati, Ohio, 1990–91

McCord Middle School, Worthington, Ohio, 1991–92

Perry Middle School, Worthington, Ohio, 1991–92

Pleasant Run Middle School, Cincinnati, Ohio, 1990–92

Coordinated by Glenn Markle (Site Coordinator) and Cynthia Geer (Graduate Assistant) based at University of Cincinnati, Cincinnati, Ohio.

Secondary Site Centers and Affiliated Schools

Arizona
Lee Kornegay Junior High School, Miami, Arizona, 1991–92

Tso Ho Tso Middle School, Fort Defiance, Arizona, 1991–92

Williams Middle School, Williams, Arizona, 1991–92

Coordinated by Diane Ebert-May (Site Coordinator) and Alison Graber (Graduate Assistant) based at Northern Arizona University, Flagstaff, Arizona.

California
Hollenbeck Middle School, Los Angeles, California, 1990–91

Coordinated by Andrea Gombar based at Los Angeles Unified School District, Los Angeles, California.

Colorado
Bookcliff Middle School, Grand Junction, Colorado, 1991–92
East Middle School, Grand Junction, Colorado, 1991–92
Fruita Middle School, Grand Junction, Colorado, 1991–92
Orchard Mesa Middle School, Grand Junction, Colorado, 1991–92
Mount Garfield Middle School, Grand Junction, Colorado, 1991–92
West Middle School, Grand Junction, Colorado, 1991–92
Coordinated by Kathleen Kain (Site Coordinator) and Rebecca Johnson (Field-Test Teacher) based at Mesa County Schools, Grand Junction, Colorado.

Michigan
Isaac E. Crary Middle School, Waterford, Michigan, 1990–92
Detroit Country Day School, Birmingham, Michigan, 1990–92
Stevens T. Mason Middle School, Waterford, Michigan, 1990–92

John D. Pierce Middle School, Waterford, Michigan, 1990–92
Coordinated by Rochelle Rubin based at the Instructional Materials Center, Waterford, Michigan, and David Housel based at Waterford Public Schools, Waterford, Michigan.

Missouri
Academy of Arts & Sciences, Kansas City, Missouri, 1990–91
Coordinated by Francesca Mollura based at the Academy of Arts & Sciences, Kansas City, Missouri.

North Carolina
Farmville Middle School, Farmville, North Carolina, 1991–92
Coordinated by Brenda Evans based at the Department of Public Instruction, Raleigh, North Carolina.

Pennsylvania
Davis School at IUP, Indiana, Pennsylvania, 1991–92
Freeport Junior High School, Freeport, Pennsylvania, 1990–92

Milton Hershey School, Hershey, Pennsylvania, 1991–92
North Hills Junior High School, Pittsburgh, Pennsylvania, 1991–92
Coordinated by Thomas Lord (Site Coordinator) and Terry Peard (Assistant) based at Indiana University of Pennsylvania, Indiana, Pennsylvania.

Wisconsin
Lundahl Junior High, Crystal Lake, Illinois, 1991–92
North Junior High, Crystal Lake, Illinois, 1991–92
Richfield Senior High, Richfield, Minnesota, 1991–92
Wilbur Wright Middle School, Milwaukee, Wisconsin, 1990–92
Coordinated by Jean Moon (Site Coordinator, 1989–90) and Craig Berg (Site Coordinator, 1991–92) and Tracy Posnanski (Graduate Assistant) based at University of Wisconsin-Milwaukee, Milwaukee, Wisconsin.

Program Reviewers

Michael R. Abraham, *University of Oklahoma, Norman, Oklahoma* (Science Content, Instructional Model)
Thomas Anderson, *University of Illinois, Champaign-Urbana, Illinois* (Reading)
Albert A. Bartlett, Professor Emeritus, *University of Colorado, Boulder, Colorado* (Science Content)
Clyde R. Burnett, *Fritz Peak Observatory, Rollinsville, Colorado* (Science Content)
Elizabeth Beaver Burnett, *Fritz Peak Observatory, Rollinsville, Colorado* (Science Content)
Kallene Casias, *Turman Elementary School, Colorado Springs, Colorado* (Cooperative Learning)
Audrey Champagne, *SUNY, Albany, New York* (Instructional Model)
Aileen Dickey, *Wildflower Elementary School, Colorado Springs, Colorado* (Cooperative Learning)
Peter Drotman, *Centers for Disease Control, Chamblee, Georgia* (Science Content)

Richard A. Duschl, *University of Pittsburgh, Pittsburgh, Pennsylvania* (Nature of Science, Science Content)
Diane Ebert-May, *Northern Arizona University, Flagstaff, Arizona* (Science Content)
Timothy Falls, *Meadows Elementary School, Novi, Michigan* (Safety)
Robert J. Francis, *GM Hughes Electronics, Los Angeles, California* (Science Content)
Terry Gerbstadt, *KRDO, Channel 13, Colorado Springs, Colorado* (Science Content)
Jerald Harder, *Aeronomy Laboratory, National Oceanic and Atmospheric Administration, Boulder, Colorado* (Science Content)
Henry Heikkinen, *University of Northern Colorado, Greeley, Colorado* (Science Content)
Werner Heim, *Colorado College, Colorado Springs, Colorado* (Science Content)
Jane Heinze-Fry, *Cornell University, Ithaca, New York* (Science Content)

Sheryl Hobbs, *Carmel Middle School, Colorado Springs, Colorado* (Cooperative Learning)
Martin Hudson, *Hughes Aircraft, Denver, Colorado* (Science Content)
Jack Lochhead, *Ventures in Education, New York, New York* (Instructional Model)
James McClurg, *University of Wyoming, Laramie, Wyoming* (Science Content)
Joseph D. McInerney, *BSCS, Colorado Springs, Colorado* (Science Content)
Verjanis Peoples, *Grambling University, Grambling, Louisiana* (Equity)
E. Joseph Piel, *Professor Emeritus, SUNY, Stony Brook, New York* (Science Content)
Belinda Rossiter, *Baylor College of Medicine, Houston, Texas* (Science Content)
Kathleen Roth, *Michigan State University, East Lansing, Michigan* (Instructional Model)
Frank Tallentire, *Aerospace Engineer, Retired, Littleton, Colorado* (Science Content)
Lynn Williams, *University of Oklahoma, Norman, Oklahoma* (Nature of Science)

Artists and Photographers for Field-Test Editions

Susan Bartel
Carlye Calvin
John D. Cunningham
Michelle Dinan
Carmen Franco-Stephenson
Suzanne Guthrie
Sandy Keller
John McDowell
Staff Photographs, National Center for Atmospheric Research
Jacqueline Ott-Rogers
Ed Reshke
Nancy Smalls
Bob Trochim
Linn Winsted Trochim

Other BSCS Staff Contributing to the Project

Cindy Anderson
Debra Hannigan
Michael R. Hannigan
Sandy Keller
Joseph D. McInerney
Jean P. Milani
Dee Miller
Dee Nolan
Carolyn O'Steen
Judy Rasmussen
Bruce Thompson
Pam Thompson
Katherine A. Winternitz
M. Jean Young

Coordination, Text Design, Electronic Production and Prepress   PC&F, Inc.

Public Support   National Science Foundation

Private Support

Science Kit & Boreal Laboratories, Inc., Tonawanda, New York   IBM Educational Systems, Atlanta, Georgia

Index

A

Advertising, 298
Africa, 218
  drought in, 206
Air mass, 209, 212, 214
Alaska, volcanoes in, 196
Al-Biruni, 122–126, 132–133, 139
Al-Khwarizmi, 123
Ancient history, 10
Animals, and garbage, 263
Arabia, 122–123
Aristotle, 116–119, 125, 147
Ash, 274–275
Asia Minor, 113
Atmosphere, garbage in, 293
Audubon Society, 298
Australia, 204–205
Average, 341

B

Bags
  paper, 297–298
  plastic, 296–297
Bar graph, 313
Beaches, and garbage, 263
Beijing, China, 119–121
Benefits, 235–236
  of composting, 276–277
  of fires, 203
  of recycling, 275–276
Birth defects, 273
Black, Joseph, 126–127
Bones, structure of, 22
Botanists, 37
Buoyant force, 187

C

Caesar, Julius, 10
Calendars, 10, 22
California, 206
  and earthquakes, 122, 155–156, 197
Careers, in science, 36–37
Cause, 41–43
Cause and effect, 41–43
Change, patterns of, 80–81
Chang Heng, 120–122, 126, 132
Chemicals, 273
Chicago, 205
China, 119–121
  drought in, 206
Chinese people, and magnets, 56–57
Cholera, 85–86
Climate, 145
Clouds, 192
Collection tanks, 274
Colorado, 199, 208, 212
Combustion, 202
Compass, 56–57
Composting, 276–277
Condensation, 187
Continental drift, 129–130, 145
Continents, 143–144
  patterns in, 144–146
  structure of, 154–155
Controlled experiments, 37–39, 229–230
Convection box, 181, 185–186
Convection cell, 185, 190
Convection currents, 203
Core, of the earth, 151
Coriolis effect, 190, 209
Correlations, 21–22, 42–43
Costs, 235–236
Cowpox, 84–85
Crust, of the earth, 150–151
Cycle, 22
  of water, 184

Cycles, 21–22
  of the moon, 70
  of trees, 35–36

D

Data table, 318
Daylight, measuring, 123–124
Decisions, 242
  balanced, 238–239
  and probability, 248–249, 254
  and problem solving, 234–236
Disasters, natural, 196–199
  probability of, 252
Doldrums, 191
Droughts, 198, 205–207
  probability of, 255
Dumps, 262
Dust storms, 218

E

Earth
  rotation of, 188
  structure of, 150–151, 153–156
  view from, 75–77
Earthquakes, 109, 151
  in California, 122, 155–156, 197
  causes of, 155–156
  detecting, 120–122
  patterns of, 135, 137, 145–146
  probability of, 255
  theories of, 114–115, 117–122
  and theory of plate tectonics, 147, 149
  and volcanoes, 147
Eclipse, of the moon, 76–77
Edinburgh, Scotland, 126
Effect, 41–43
Egyptians, 10, 15–16
Eismitte, Greenland, 129–130
Electricity, and garbage, 275
Energy, and recycling, 275
England
  cholera in, 85–86
  compass in, 57
Environment, 297
  and drought, 206
  products and, 296
Epidemic, 85–86
Europe, 268

Evaporation, 187
Evidence, 101–103
  gathering, 64
  organizing, 132–133
Experiments, controlled, 37–39, 229–230
Explanations, 103
Explanations, scientific, 72, 75–77, 132
  value of, 84–86
Eye, of hurricane, 210

F

Factors, 36, 40
Farr, William, 85–86
Faults, 158–159
Fire lines, 202–203
Fires, 198, 202–205
  and drought, 206
  in London, 216
  patterns of, 204
  preventing, 200
  probability of, 252
Fire triangle, 202
Flash floods, 208
Floodplain, 207–208
Floods, 197, 207–208
  and hurricanes, 211
  patterns of, 207
  probability of, 207
Floral designers, 37
Folding, 158–159
Fossils, 145
France, 128

G

Garbage, 261, 285
  composting, 276–277
  crisis in, 270–277
  disposal of, 261, 264, 272–273, 288
  and electricity, 275
  and health hazards, 263, 273–274
  incineration of, 274–275, 293
  and patterns, 264
  and plastics, 263
  in U.S., 262, 264, 271–272, 285, 289
  and U.S. population, 277
Geologists, 145–146
Geology, modern, 126–130
Gilbert, William, 57

Global winds, 187, 191
Graphs, 14, 45, 51, 87
    bar, 313
    line, 325
    of population, 27
Greece, ancient, 116–119
Greenhouse worker, 37
Greenland, 129–130
Groundwater, 273
Growth
    patterns of, 42
    of trees, 35–36

H

Hadley, George, 187
Hadley cell, 187
Hawaii, 161
    volcanoes in, 159, 160, 196
Hazardous waste, 273–274
Health, hazards to
    and garbage, 263, 273–274
Heat energy, 184–185
HGH, 42
Horizontal axis, 305
Horse latitudes, 191
Hot spots, 160–161
Houses, 231–233
Human growth hormone (HGH), 42
Humidity, and fires, 203
Hurricanes, 59, 209–211, 212, 219
    Camille, 211
    data on, 220
    Dora, 59
    Emily, 225
    Gabrielle, 221
    Gilbert, 22
    Helene, 23
    Hugo, 197, 221–222
    Joan, 223–224
    patterns of, 252
    probability of, 252
Hutton, James, 126–128, 132

I

If–then statements, 104
Igneous rocks, 159–160
Illness, patterns of, 84–86, 282–284
Incineration, 274–275, 293

Information, quality and quantity of, 55–56, 63, 72, 235
Island of Malta, 60
Italy, 125

J

Japan, 154, 159
Julian calendar, 10
Jupiter, 168

L

Landfills, 261, 280, 298
Lava, 159–160, 196
Lead poisoning, 274
Leakey, Louis, 22
Leakey, Mary, 22
Legends, about the moon, 66–67
Limits, of recycling, 276
Lisbon, Portugal, 109
Litter, 263, 289
Lodestone, 56–57
London, 85–86
    fire in, 216
Love Canal, New York, 273
Lower mantle, of the earth, 150–151

M

Magma, 154–155, 159–160
Magnetic fields, 57
Magnetite, 57
Magnets, 52, 55–57
Mantle, of the earth, 150–151, 161
Mapmakers, 123–124, 139
Maps, 128
    ancient, 122–125
    of ocean floor, 153
    patterns on, 133
    as tools, 164
    world, 139
Mars, 168
Measurements, 10–11, 25
    of daylight, 123–124
Metamorphic rocks, 160
Microbe, soil, 276
Middle Ages, 268
Miletus, Turkey, 113
Missouri, 199
Moon, 24

legends of, 66–67
patterns of, 24
phases of, 70, 72, 75–76
Moro, Abbe Anton–Lazzaro, 125–126
Mountains, 125
    formation of, 158–159
    on ocean floor, 153–154
    volcanic, 159–161, 168
Mt. St. Helens, 196

N

National Park Service, 200, 204
Natural disasters, 196–199
Nature, patterns in, 22
Neptune, 170
New York, 273
Number scale, 306

O

Ocean floor, 134, 136, 153–156
Options, 236
Oral tradition, 122
Outer space, 163, 168–170
    volcanoes in, 168

P

Packaging, reducing, 276
Pangaea, 148
Paris, France, 128
Patterns, 13–16, 282
    of change, 80–81
    in continents, 144–145
    of earthquakes, 135, 137, 164
    of fires, 204
    of floods, 207
    and garbage, 261, 264, 288
    of growth, 42
    of hurricanes, 252
    and legends, 66–68
    and the moon, 24, 66–67
    in music, 25
    and natural events, 254
    in nature, 23
    on maps, 132–133
    on ocean floor, 134, 136
    and predictions, 13–16, 52–53
    and puzzles, 27–29

recognizing, 80–81
    of rocks on earth, 135, 138
    of tornadoes, 252
    and trees, 35–36
    types of, 21–22
    value of, 84–85
    of volcanoes, 134–135, 137, 164
    of weather, 184–185, 187–188, 217
    of winds, 187–188, 190–191
Pendulum, 16, 21–22
Phases, of the moon, 70, 72, 75–77
Plagiarism, 336
Planets
    Earth, 150
    Jupiter, 168
    Mars, 168
    Neptune, 170
    Pluto, 170
    Venus, 163, 169
Plants, growth of, 32–41
Plastics, 263, 270–271
Plates, of the earth, 146–147, 151, 160
Plate tectonics
    and earthquakes, 147, 149
    theory of, 146–147, 149, 151, 164
    and volcanoes, 147, 149
Playfair, John, 127
Pluto, 170
Poisoning, lead, 274
Pollution, 255, 275, 280
    products and, 296
    water, 85–86, 273
Pope Gregory XIII, 10
Population, in U.S., 277
Predictions, 63
    accurate, 55–56
    and patterns, 13–16, 52–53
    and scientific explanations, 72
    and theory of plate tectonics, 147
    weather, 59–60
Prehistory, 269–270
Probability
    of natural disasters, 254–255
    statements of, 247–248
Problems, solving, 233, 235
Products, recycled, 276, 285, 287
Propellants, 293

Puzzles, 15–16, 97–100
    and patterns, 27–29

Q

Quality, of information, 55–56, 63, 72
Quantity, of information, 55–56, 63, 72
Questions, scientific, 101

R

Rain, 197, 207
    and landfills, 272, 280
Recycling, 270, 275–276, 285, 287, 291–292
    arrows, 299
    benefits of, 275–276
    closed–loop, 299
    limits of, 276
    of plastic bags, 296–297
Reservoirs, 208
Resource, tires as, 287
Resource list, 294
Ridges, on ocean floor, 153–156
Risks, 235
Rivers, 208
Rocks
    ages of, 135, 138
    formation of, 159–160
Rosetta stone, 15–16
Rotation, of the earth, 188
Round off, 302

S

Sahara Desert, 218
St. Louis, Missouri, 296–299
San Andreas fault, 155
San Diego, California, 296–299
Saudi Arabia, 123
Scientific explanations, 72, 75–77, 132
    value of, 84–86
Scientific questions, 101
Scientists, 103–104
    dilemmas of, 103
    and problem solving, 282–285
Scotland, 126
Sea lions, 263
Sedimentary rocks, 160
Smallpox, 84–85
Smokey the Bear, 200

Snider, Antonio, 128, 130
Snow, John, 85–86
Snowstorms, probability of, 252
Soil microbes, 276
Solar system, 169–170
South Carolina, 197
Statements of probability, 247–248
Strabo, 118–119, 132

T

Tambora, 109
Technology, 233
Tectonics, plate, 146–147, 149, 151, 164
Temperature, and fires, 203
Tests, 103
    fair, 229–230
Texas, 211
Thales, 113–115, 132
Theory of plate tectonics, 146–147, 149, 151, 164
Thunderstorms, 207–210, 212–214
Time, measuring, 10
Tires, automobile, disposal of, 287
Tools, 164
Tornadoes, 199, 211–215, 217
    patterns of, 252
Toxic waste, 273–274
Transform faults, 155–156
Trash, 285
Trees, cycles of, 35–36
Trenches, 154–155
Trends, 21–22
Turkey, 113
Twisters, 218

U

United States
    garbage in, 262, 264, 271–272, 285, 289
    population in, 277
United States Weather Service, 59
Upper Mantle, of the earth, 150

V

Venus, 163, 169
Vertical axis, 305
Volcanoes, 109, 151
    in Alaska, 196

and earthquakes, 147
in Hawaii, 160, 196
in outer space, 168
patterns of, 134–135, 137, 145–146
probability of, 252
theories of, 117–119, 125, 127
and theory of plate tectonics, 147, 149

W

Waste, hazardous or toxic, 273–274
Water, 274
    ground, 273
    in landfills, 272–273
    movement of, 187–188

Water cycle, 184
Weather
    and fires, 203
    forecasting, 247
    patterns of, 184–185, 187–188
    predictions about, 59–60
Wegener, Alfred, 129–130, 145
Wind, 184–185
    and drought, 105
    patterns of, 187–188, 190–191
    and tornadoes, 213–214

Y

Yellowstone National Park, 161, 198, 204